Ozark Baptizings, Hangings, and Other Diversions

Ozark Baptizings, Hangings, and Other Diversions

THEATRICAL FOLKWAYS OF RURAL MISSOURI, 1885-1910

by Robert K. Gilmore

FOREWORD BY ROBERT FLANDERS

UNIVERSITY OF OKLAHOMA PRESS : NORMAN AND LONDON

Library of Congress Cataloging-in-Publication Data

Gilmore, Robert K. (Robert Karl), 1927–
  Ozark baptizings, hangings, and other diversions.

  Bibliography: p. 253
  Includes index.
  1. Ozark Mountains Region—Social life and customs. 2. Amuse-
ments—Ozark Mountains Region—History. I. Title.
F472.09G55   1984   977.8'835   83–40324
ISBN 0–8061–2270–6

The illustration on pages ii–iii shows the Jacob Goetz farm, Taney
County, 1907. Courtesy Museum of the Ozarks.

4   5   6   7   8   9   10   11   12   13

*To the wonderful old-time Ozarkers
who so graciously shared with me
their time and their memories,
this book is affectionately dedicated.*

Contents

Illustrations

xi

MAP

Foreword

BY ROBERT FLANDERS

THIS is a book about the Ozarks, and about Ozarkers. To understand the peculiar relation between the region and its people is to understand that neither would exist without the other.

The Ozarks is a distinctive highland region of southern Missouri, northern Arkansas, and eastern Oklahoma. It has been historically an isolated region, but only *relatively* isolated in comparison with other surrounding regions like the prairie-plains of Kansas or the Missouri and Arkansas River valleys. The region is internally variegated, with forest and prairie, gentle slopes and steep, fair soils and poor. The people, until the onslaught of the twentieth century, have *not* been as variegated. Overwhelmingly white and Protestant with a Scotch-Irish majority and German and English minorities, they have had an ethnic and cultural cast relatively homogeneous when compared, say, with Kansas and its great ethnic diversity. The people have been culturally isolated, not only because the Ozarks was a difficult place to get in or out of or through but also because they were isolated in history from those great cultural mainstreams of Western civilization of the last half millennium, the Renaissance, the Enlightenment, and modernity.

Baptizings, Hangings, and Other Diversions is about a single generation, from 1885 to 1910, of Ozark history. The period is felicitous for the author's purpose because it was the generation of remembered childhood and youth for his elderly informants. And it was certainly a period that presented an unusual pithiness in county and small-town newspaper writing, Gilmore's other major source. I suspect that that turn-of-the-century generation was one that enjoyed something of an efflorescence in rural and small-town culture. The particular horror of the Civil War was sinking into history. Railroads were still being built, with all the anticipation, optimism, excitement, and new experience that accompanied those portentous events. New people swelled the population, including new kinds of people, especially Yankees and townsmen and women, who knew from experience what towns were and how one lived in them. New church denominations arrived, with their programs and their organizations; public schools arrived in the same way. Indeed, much of the organizational and technical paraphernalia of modernity moved in. But the old ways, the old farms, roads, and buildings, the old people were not swept away or overwhelmed at once. The new and old coexisted in exuberant tension.

A root of the ironic style of humor so well employed by the Ozarkers may well have proceeded from that tension. The encounter of the more modern with the less modern, of the newfangled with the oldfashioned, contained in their potential for misunderstanding and conflict the prized elements of irony. The debates of the Friday-night literary described in these pages reveal both conflict and humor. Women's rights; education (by which was always meant schooling); new modern religious ideas; new notions of social control; money, that complicated and, to many, new abstraction; the new corporate organization —all these and more were represented in topics that were debated. The resolutions, lacking the prolix and obfuscating language of modern social science to which we have become accustomed, seem quaint and old-fashioned in their disarming directness, as the author's examples attest. The classic and ultimate response was that of a correspondent reporting the debate "Resolved: That life is not worth living," which, being decided by the audience in favor of the affirmative, "has made us feel so bad we wasn't able to eat but seven biscuits for breakfast."

The period ending the nineteenth century and beginning the twentieth was an interstice in Ozark history, a brief period when modernity seemed to settle with apparent benignity upon old traditions. It was the close of a time without beginning when primary, face-to-face relationships and oral communication formed the essential character of society. It was a relatively prosperous, secure, felicitous time. The solutions to primordial difficulties of living—ignorance, scarcity, unremitting toil, distance, loneliness—seemed at hand. Railroads; more towns with their amenities; churches with temporal as well as eternal salvation; writing, usually in newspapers; and the ability to read, gained in the new public schools, were realities. The twentieth century was a welcome, hopeful century at its outset. The old was not abruptly replaced or obliterated by the new. Rather, there was a melding, a flowing together, which was incongruous but not infelicitous. The irony of the incongruity has been captured with singular success in the pages that follow. The oral tradition, passing into the programs and scheduled events of the new institutions, organizations, and societies, was given further verbal styling in the reportage of sensitive editors and "neighborhood correspondents," whose writing as here preserved has elements of provincial literary genius. Those

Ozarkers, inclined to be dramatic rather than phlegmatic, articulate rather than taciturn, especially appreciative of irony and skillful in its use, turned the portentous cultural shifts of their time into an ever-flowing spring of genial humor. Thoughtful individuals among them—and there were many—appreciated the drama of change all about them, and, whether amazed, alarmed, amused, or infuriated by it, they found it always interesting. *Ozark Baptizings, Hangings, and Other Diversions* provides a unique window on a brief golden age in the Ozarks, before the verdict of the twentieth century had been returned.

Now, almost a century later, the Ozarks has experienced histories scarcely imaginable to the generation of this book. Hardship became grinding poverty. A spare land was despoiled by mining, timbering, floods, erosion, overpopulation, and ignorant overuse of the fragile ecosystem. The still-lush wilderness of forest and prairie of the nineteenth-century Ozarks gave way to a wasteland of the twentieth. Then the Ozarks changed its face again. Conservation restored forests, restrained floods, restored watersheds, preserved precious soil, and created great reservoirs that generated both electricity and tourism. The Ozarks is becoming cosmopolitan, as everywhere else. But beneath the surface of the newness, many of the old ways, the old value systems, the old social and personal attributes persist. Among the survivals is the oral tradition, the love of the spoken language. I have observed it repeatedly in the shooting of documentary films in which I have recently been engaged. One octogenarian farmer, saying, "Well, the woman says I've got to show you my pig, and I sure want to do that!" led the film crew over to a little pen. A brief search revealed a ridiculously small, pinkish-white runt piglet, scarcely tall enough to see over the edge of its dish. It looked up at the camera inquiringly and emitted a weak, puckish grunt. The farmer, pushing open the little gate of the pen, said, with just the faintest trace of matching puckishness, "Now, a man has to know his pigs to go in with *that!*"

ROBERT FLANDERS
Director, Center for Ozarks Studies
Southwest Missouri State University
Springfield

Preface

THIS book celebrates the centennial of an era in which
a generation of people just emerging from a frontier way
of life found that it had the time, the opportunity, and the
desire to provide itself and its children with an improved
quality of life—cultural activities, education, expanded
social contacts, and entertainment. Their efforts were sometimes crude and often amusing to those of us who live in
a more "sophisticated" age, but the results left a mark
upon the singular character of the place that is the Ozarks
and upon the people who call the Ozarks home.

My concern with the subject comes from a melding of
my interest in theater with my love of the Ozark hills
and its people. I grew up on an Ozark farm and attended
a one-room country school and enjoyed Friday-night literaries, pie suppers, and spelling bees there. My grandparents, who reared me, were of the generation described
in this book, and most of the attitudes and practices of the
community I knew in the Ozarks of the 1920s and 1930s
were essentially the same as those of fifty years before.
Almost all of the people I knew, both of my own generation and that of my grandparents, had always lived in the

On the road near Thayer, Oregon County, early 1900s. Courtesy Lurine Bryan and the *West Plains Gazette*.

Ozarks, most of them within a few miles of their present homes.

This sense of home, of belonging to a particular place, has always been strong among Ozarkers. I have imagined that the first settlers from Kentucky and Tennessee saw in this isolated, protected land exactly the opportunities for the kind of life they wanted to live and somehow imbued their descendants with this sense of rightness of place.

Not long ago several young people of Shannon County, an Ozark county adjacent to the region included in this book, were interviewed for a documentary film. When they were asked whether they would leave Shannon County when they finished school, many answered yes, but with many expressions of sadness and regret. There simply would not be jobs for them, they felt, and their leaving would be not by their own choice but out of economic necessity.

Ozarkers are rather sensitive to whether one is a native or a "come here," as a friend of mine, himself a native, calls a newcomer. Some years ago I was interviewing, in the company of two old-family Ozarks sisters, a man of sixty-five or seventy years who was giving me, I thought, much good information and insight about the area. When he left the room for a moment to answer a phone call, the ladies watched him go, after which one of them sniffed and said indignantly to her sister, "He always passes himself off as an Ozarker, but you know his family came here from St. Louis when he was four years old." The fact that this well-liked and highly respected "come here" was of more than sixty years' residence did not qualify him to talk with authority about his adopted land.

The popular mythology about the early Ozarkers would have them a suspicious lot (never trust a stranger), fiercely independent (never be beholden to any man), and cherishing their solitude (why else would anyone choose to live miles from the nearest neighbor?). These characteristics are, I believe, generally true of Ozarkers, both of today and of the past. But like most other generalizations they describe only one aspect of a complex, multifaceted personality.

Suspicious? Yes, we are reserved and cautious when we are approached by someone we do not know, or when a new idea or way of doing something is advanced. But our suspicions are matched by an openness and generosity toward a friend or toward a stranger in need.

Independent? A man needs to be able to do for himself, for to rely too much on others is to be in their debt, and indebtedness endangers freedom. But being a neighbor means more than simply living on the next farm. When many hands are needed, as at harvesttime or in time of sickness or sorrow, we both give and gratefully accept help without reservation. We are part of a community, and it is our privilege and joy to be proud participants in the life of that community.

Solitary? Many of us just do not like to live where too many people push in on us. We cherish the right to be alone, yet even those of us who cannot abide hearing our neighbor's ax nevertheless know that he is there, over the ridge, and we are glad. We seize eagerly the opportunity to be together with our friends and our neighbors on sociable occasions, for loneliness is hurtful; companionship, refreshing.

Therefore, it is not surprising that the Ozarkers of an earlier time got together on many occasions for recreation. I do not find it extraordinary that they seemed particularly to enjoy those entertainments in which they could find, or insert, dramatic or theatrical elements. Throughout history human beings have used forms of art, including drama, to help them meet their need to understand better the world in which they live; to communicate to others their intellectual ideas and emotional feelings; and to realize a vision of themselves and their lives that is superior

A family gathering at a shady spring on an Ozark farm, 1902.
Courtesy Museum of the Ozarks, Springfield.

Ozark fishermen displaying their catch on the Gasconade River, about 1900. Courtesy Museum of the Ozarks.

to the realities of the narrow borders of their immediate surroundings. It would be a mistake to describe as no more than diversions the many entertainments with theatrical qualities enjoyed by the Ozarkers. They found in many of these events opportunities for communication, self-realization, and enlightenment—opportunities that were otherwise denied them or made more difficult because of the circumstances in which they lived.

The instinctive craving for drama never completely dies out, but is kept alive through the determination, ingenuity, or machinations of a people to whom theatrical activity of some sort is highly desirable. Theater has struggled to live, and has lived, in spite of the most unfavorable conditions, through performances of wandering minstrels, jugglers, and acrobats; in the crude tropes of the medieval church; and disguised as interludes to musical entertainments or "moral lectures." The Ozarkers found still another means of preservation—theatrical folk entertainments.

ROBERT K. GILMORE

Sources
and Acknowledgments

MATERIAL for this book was drawn from two major primary sources: scores of personal interviews with settlers who lived in the Missouri Ozarks between 1885 and 1910 and weekly Ozarks newspapers published during that time. The interviews and the newspapers came from a twelve-county area* centered around the famous Shepherd of the Hills and White River country, an area that in modern times styles itself the "Heart of the Ozarks" and which, indeed, may be representative, both geographically and culturally, of the greater Ozarks of which it is the core.

The editors of the weekly newspapers relied heavily on volunteer country correspondents, who wrote strongly expressed and highly biased opinions, often phrased in uniquely ungrammatical form, all of which added color and depth to the information they presented:

> We have received from a minister in Eureka Springs an offer to hold a meeting here. Think of it, brethren, a minister offering to preach for us in this time of scarcity. For years we have been left dur-

*The twelve counties are Barry, Christian, Douglas, Greene, Howell, Lawrence, Ozark, Stone, Taney, Texas, Webster, and Wright.

ing the winter time to wander in heathen darkness, and only when spring chickens, sweet potatoes, and other luxuries were the bill of fare could we hope for any ray of light. [*Cassville Republican*, February 23, 1899]

Some parties came to this place Monday selling and advertising a patent medicine. They erected a stand between Gilmore and Bailey's furniture store and the new hotel from which place they gave a free minstrel entertainment Tuesday evening. During the progress of their entertainment a volley of ancient eggs were sent into their midst. Several parties being struck. The guilty parties who perpetrated the dirty trick should be dealt with unwashed hands if found out. The egg business has went to an entire extreme in Ash Grove and this extravagant use of them should be suppressed. [*Ash Grove Commonwealth*, October 14, 1886]

In these choice quotations spellings and punctuation have been left as I found them.

Most of the interviews were conducted in 1960. The old-time Ozarkers recalled for me personal experiences connected with the entertainments of their youth, and many vividly recalled instances from very early childhood. Mrs. Eva Dunlap, of Ash Grove, for example, described with considerable clarity an event that occurred in 1882, when she was six years old. In celebration of her birthday her father erected a temporary stage on which she and other members of her family gave recitations for a small audience of neighbors. When her turn came, she was pushed protesting to the platform by her mother's firm hand. There, standing under a wagon bow entwined with wild roses, she recited her "piece": "Comb my hair straight, Sister Janie./Gather all the curls away./For they lie so hot and heavy/On my burning cheek today./Hold me close and kiss me, Mother./Do not cry and tremble so./ I'm very tired, Mother,/And I do not fear to go."

The interviews were recorded, and transcriptions of many of them are included in the appendix to this book. Most of these old residents have passed away, and with them has gone an invaluable and irretrievable store of memories, except as they have been captured in this book. The opportunity to record this particular segment of a fast-disappearing culture has been a source of great personal satisfaction to me. To each of the great old-timers to whom I talked and who received me and my tape recorder with true Ozark hospitality, I extend my special thanks. Thanks also to Elias Johnson and Nathan King for help with the map; to my daughter, Julie Bloodworth, who typed the manuscript; to my wife, Martha, an excellent critic and editor; and to the very helpful staff of the State Historical Society of Missouri. Although many, many others helped me as this book was being developed, I particularly wish to acknowledge the assistance and encouragement of three very special friends, Milt Rafferty, Bob Flanders, and Gordon McCann.

Anyone interested in the Ozarks, past and present, should not fail to examine the *Missouri Historical Review,* the distinguished quarterly of the State Historical Society of Missouri; the *Ozarks Mountaineer,* an interesting magazine of varied topics; the bimonthly *West Plains Gazette,* a handsome magazine with many outstanding photographs in each issue; and that delightful magazine *Bittersweet,* which, until it recently ceased publication, was the loving creation of Lebanon High School students. In addition, much good information can be found in many of the Centennial publications produced in recent years by several Ozark communities. Some county historical societies publish excellent journals, among them the *White River Valley Historical Quarterly.*

RKG

Ozark
Baptizings,
Hangings,
and Other
Diversions

Introduction

T HIS book is about the people of the Missouri Ozarks and
how they entertained themselves and were entertained by
others during one particular generation, a twenty-five-
year span beginning about 1885.

The end of the nineteenth century was an exciting
time for America. It was a time of change, of self-improve-
ment, and of discovery. The nation believed that it had
reached a certain maturity, for, after all, it had celebrated
the centennial of its founding. Of course, as nations go,
the United States was but an adolescent, so coltish exu-
berance won out over blasé sophistication as a national
attitude, and the country struggled to absorb the wonders
of a ten-story skyscraper, electric street railways, tele-
phones, talking machines, fantastic industrial growth, and
the many mechanical marvels of the age. The nation was
joined East to West by a transcontinental railroad. The
Civil War was behind, and its wounds were healing. In
1885 there were thirty-eight states in the Union. Missouri
had been a state since 1821, its entry with Maine a key
ingredient in the Missouri Compromise, by which a balance of slave and free states was maintained as the Union
expanded.

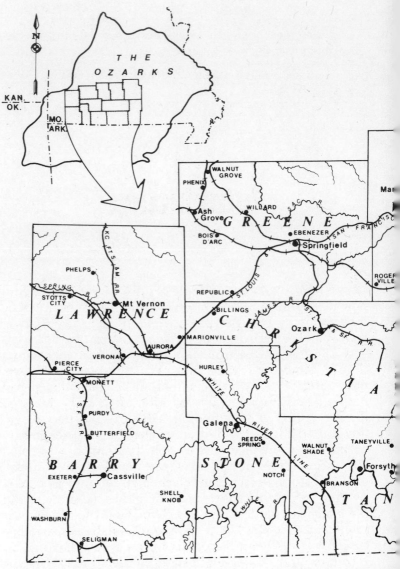

source: The Missouri Railway and Warehouse Commission Official M
J. Bien & Co. Lith. N.Y.

cartographer: Nathan King

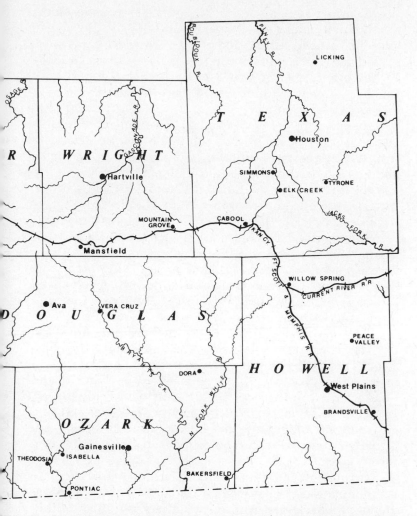

The Ozarks in 1904. Based on Missouri Railway and Warehouse Official Map, 1904, J. Bien & Co., Lithographers, New York City, Nathan
King, cartographer.

The first settlements in Missouri were in the plains areas, chiefly along the Missouri and Mississippi rivers. The plateau-and-hill regions of the central Ozarks were settled late. In his famous trip through the Ozarks in 1818–19, the ethnologist Henry Rowe Schoolcraft found only small settlements and individual families, mostly those of hunters and trappers.[1]

Even before the Civil War larger numbers of settlers had begun to appear in the region. They were primarily from Tennessee and Kentucky, hill people of Scotch-Irish and English descent who found in the Ozarks a way of life and an isolation that compared with what they had left. The absence of competition from slaveholders and large-scale farmers made the region particularly attractive.

After the war the population increased as a flood of immigrants poured in. These new settlers included Yankees as well as southern highlanders, and with them came new ideas and the embodiments of modernity: newspapers, railroads, public schools, denominational religion, and larger and more complex social organizations.

The new never quite overcame the old, however. Perhaps it was the isolation of the Ozarks and the character of the earliest settlers that contributed to the creation of a culture that remained remarkably unchanged and unaffected by influences of the outside world. Robert Flanders calls it a "perpetuated frontier culture" and suggests that the older ways not only continued to exist but actually interacted with the new ways in a culturally complex fashion.[2]

Because of the influence of this "frontier culture" there is a temptation to characterize the Ozarks of the 1880s as quaint and backward, and the Ozarkers as doltish hillbillies. Such a view has a certain comfortable charm and conforms to the stereotypical image so carefully nurtured in recent years by motion pictures, television, and the tourist industry.

6

In truth the Ozarks of the exciting decades of the turn of the century was still very much a part of the United States, and the interests and aspirations of its citizens were no different, except perhaps in degree and opportunity of fulfillment, from those that could be generalized about the rest of the nation. Ozarkers had free public schools, places of worship (many of them), opera houses, newspapers, a strong sense of patriotism, community pride, arguments over free silver, McGuffey readers, railroads, and mail-order catalogs. The old isolation was breaking down. There were intellectual and cultural activities elsewhere in the United States, and the Ozarkers knew of them and wanted to share in them.

Although life was hard for most rural families, survival no longer demanded continuous, hardscrabble, unrelenting labor. Time remained, at least at certain seasons of the year, for even the most remote farm family to participate in special entertainment events. Through mail-order catalogs the wonders of the urban department stores were known to the Ozarks farmer and, if he could afford them, were readily available to him.

Through the weekly newspapers (fifty-five were being published in 1899 in the twelve-county area covered in this book), Ozarkers knew of the symphonies and the operas and the theater and the other cultural entertainments available to the citizens of New York and Boston and Chicago and Saint Louis. But such activities as these demand the large number of patrons and the transportation system that can be found only in an urban setting. The Ozarks was a sparsely populated, rural area, deficient in improved roads to permit patrons to attend professional entertainments.

Few traveling professional troupes penetrated the Ozarks because of the difficult terrain. The larger, more prosperous towns situated on railroad lines accommodated traveling shows, some of them repertory companies that

Three Ozark couples prepared to depart for a Sunday-afternoon gathering. They pause in their elegant buggies to have their picture taken by professional photographer O. W. Carter. Courtesy Mrs. Agnes Allen.

played as long as a week, though more commonly for one-to three-night stands. For the most part, however, the "professional entertainments" that visited the small Ozarks towns were of a variety that did not require the elaborate scenery expected of most urban theater productions. A surprising variety of performances was available. Town residents and visitors from the country were treated to several kinds of itinerant street performers. Among the entertainments mentioned in the weekly newspapers were freak shows, trained-dog acts, a hunchback "reader," a glass-and-tack eater, a strong man, a contortionist ("double back limberness"), street preachers, Montgomery Ward and Company's Advertising Show, trained bears (an "Outfit of Beggars of the Turkish Tribe"), a grand balloon ascension and tightrope walking, street musicians, and, of course, medicine shows.

Entertainments were also provided in the opera houses that stood in most of the larger towns. Representative of the shows were Major Hendershot, the Original Drummer Boy of the Rappahannock; The Edison War-Graph Company (movies); Swiss Bell Ringers; three fat sisters, a traveling dog circus; snake shows; "Sam Wright, the human frog, [who] eats glass, tacks, knife blades and live chickens"; a "short-haired dog and long-haired man"; and blind musicians playing various instruments—guitars, hand organs, fiddles, and pianos. The musicians were referred to by name, such as Blind Jasper, Blind Charley, and Blind Amos. The most popular and best known of these was "Blind Boone" (John William Boone), a black pianist who traveled throughout the United States and Europe playing a varied repertoire ranging from the classics to plantation melodies. In the Ozarks his most famous piece was "The Marshfield Cyclone," about a very destructive storm that had struck that Webster County town in 1880. It was in the opera houses that the relatively few professional touring plays were performed.

Circuses made occasional visits to the Ozarks, and citizens boarded trains to attend entertainments in Springfield, Saint Louis, and Kansas City. It took a train of nine coaches to carry passengers from Lamar, in Barton County, to a Springfield fair in 1886,[3] and the Frisco Railroad ran a $2.50 excursion train from Springfield to Saint Louis in 1895 to give the people a chance to witness the Veiled Prophet's parade. (The Veiled Prophet celebration began in Saint Louis in 1878, and was modeled after the New Orleans Mardi Gras.)[4]

Street fairs were held in many towns, especially on the Fourth of July and again in late fall, when the crops had been laid by and farmers could participate. The *Ash Grove Commonwealth* (Greene County), never modest when civic pride was involved, described the fair of 1905 as "the biggest thing that ever happened in this part of the United States. . . . It is conceded by all present that Ash Grove is at the exact center of the cosmos. The hub around which all created suns and worlds revolve."[5] The editor "conservatively" estimated the crowd on Saturday at ten thousand.

The great majority of Ozarkers, however, were denied easy access to this exciting variety of entertainment.* Most rural residents continued to rely on their own resources and those of the community to provide them with diversion from the difficult and often lonely business of everyday living.

Many rural Ozarkers were separated from the city, often even from the little village close by, physically, spiritually, and economically. Some of the roads were so bad that journeys of any distance were undertaken only in absolute

---

*In 1890 less than 19 percent of the population of the twelve counties lived in incorporated areas. If Springfield, with a population of 21,850, by far the largest city in the Missouri Ozarks, is excluded, the percentage drops to less than 9 percent. Only ten towns and villages had populations of 1,000 or more.

10

An Ash Grove (Greene County) street fair. The fairs always drew large crowds from the town and the surrounding countryside. Courtesy Geraldine T. Johnson and John Hulston.

necessity. Newspaper accounts tell repeatedly of injuries received when occupants of a wagon were thrown as a wheel passed over a stump; many of the early roads were formed by simply cutting down trees to stumps low enough for wagon wheels to pass over them. The road was then theoretically improved by the wear of travel.[6] Tom Rideout, who lived fourteen miles from West Plains, in Howell County, said that his family made the trip to town only every four or five months to get supplies. Mrs. Queen Bell said that her family did not go to town (Hartville, Wright County) very often, even though they lived only four miles away: "My dad never believed in going places unless you had business. On Saturday he usually went to town to buy what groceries we had to have. Of course, we raised most everything except coffee and sugar and things like that. On Saturday was his day for going to town."* For most farm families a trip to town to see a stage play or other entertainment and the long return journey late at night were strenuous undertakings and normally out of the question.

It is a bit more difficult to define the spiritual gulf that seemed to exist between the town and country people. Rideout said frankly, "The country people and the town people didn't pay much attention to each other. The country people had affairs of their own, and the town people did the same."

One Ozark couple who left their farm in the early 1900s to live in Kansas City, where the husband had a good job with the railroad, returned to their farm after scarcely two years. Neither of them liked city life, and besides, as the wife said, "we didn't want our daughter growing up with those city kids."

*This and following quotations that are not identified as taken from a newspaper or other written source, are from recorded personal interviews. Most of the interviews appear in full in the Appendix.

Harold Bell Wright, in his famous novel of rural Ozark life *The Shepherd of the Hills*, illustrates the view of the country people. His main character, Dad Howitt, flees from the pressures and corrupting influences of city life to the perfect peace and primeval purity of the Ozark hills. Wright extends the same idea in the lover's plot of the story by showing Ollie Stewart, who had gone to the city to work, returning a dandified snob, definitely the worse by comparison with his rival suitor, the uncorrupted stay-at-home, Young Matt, who says that he has never so much as seen a railroad. Naturally Young Matt wins the girl.

Wright also indirectly illustrates the city view when his backwoods heroine expresses fears that Ollie, having been exposed to life in the town for a time, will find her ignorant and "not a lady."

Although the rural resident recognized a "moral" distinction between the big city and the neighboring town where he did his trading, the difference was one rather of degree than of kind. He still regarded his way of life as the naturally correct and moral one and looked with some suspicion, or at least pity, on his urban neighbor. A rural correspondent wrote:

> How thankful we farmers ought to be that we live in the country instead of the noisy busy dusty city. How much nicer after the hard day's work to go out among the flowers and trees where the cool shade invites us to rest. I once lived in a large city and how I longed for a breath of sweet country air. I pity the poor little children that are shut up in the city these warm spring days.[7]

Lack of intercourse between the two groups tended to strengthen each in its particular views and in turn prevented the free social mixing that might have changed or modified them. This situation fed upon itself until modern inventions and improvements became common enough to both groups to break down the old barriers.

Another factor that maintained and even widened the gulf between the town and country folk was economic. The farmer often looked upon the merchant as a necessary evil who charged outrageous prices for goods that he knew the farmer must have. The farmer often tried to retaliate by patronizing mail-order houses and itinerant peddlers, and the merchant responded with newspaper and face-to-face campaigns in which he reminded his rural friends of the many benefits to be gained by "patronizing the home store." Debating societies discussed the topic "Resolved: That patronizing mail-order houses is detrimental to the best interest of the country," and newspapers editorialized in the following vein:

> Who sympathized with you when your little girl was sick? Was it your home merchant, or was it Sears and Roebuck? Who carried you last winter when you were out of a job and had no money? Was it Montgomery Ward and Co. or was it your home merchant? When you want to raise money for the church or for some needy person in town do you write to the Fair Store in Chicago or do you go to your home merchant? How much do Seegle Cooper and Co. give toward keeping up the sidewalks of the town or paying the minister's salary? When you were sick how many nights did Hibbert, Spencer and Parklet sit up with you? When your loved one was buried was it your home paper which shed the tears of sympathy and uttered the cheering word or was it some Chicago or New York paper?[8]

To the rural Ozarker such "benefits" often served only to deepen the mistrust and incompatibility of interest, for he was often in debt to the merchant who had both extended him credit and lent him money. The farmer hated to face the man to whom he owed money. He disliked the merchant for the simple reason that he was indebted to him and therefore avoided him as much as possible.

Not all country people were in debt to the merchants, of course, and an attempt to explain a city-country tension solely on such economic grounds would be erroneous. Other factors were certainly involved, not the least of which was the natural pride that everyone finds in his occupation, his surroundings, and his habitat. The country dweller who had little money to spend on entertainments certainly would not spend that money in town—on a foreign enterprise at that—but would choose to support his local community through activities of his schools and churches. A prideful boast by a rural correspondent in the *Ash Grove Commonwealth* of January 25, 1900, is typical of many that appeared in almost every issue of the weeklies: "The social at the [New Site] church Friday night was a financial success. A large crowd was present but order prevailed. We have as good a community as can be found in this or any other state. Our church is on a solid foundation and will stand forever if not blown away or burnt."

Perhaps, however, the prime reason that the few professional plays that did penetrate the Ozarks were not better patronized was not economic factors, distrust of the cities, or difficulties of transportation but an ethical judgment that condemned both stage plays and actors (particularly actresses) as immoral and attendance at plays as tantamount to eternal damnation. On January 13, 1887, the *Ash Grove Commonwealth* reprinted (from a Brooklyn paper) on its editorial page the following extract from a sermon by the Reverend T. Dewitt Talmadge (whose sermons were carried in full in many of the weekly papers of the region):

> What an unclean mess of theatrical stuff has been loaded upon our shores the past few years! What numbers of unclean, sensational, and reprehensible actors

15

and actresses have come here to insult respectibility. They are no more celebrated for dissoluteness than for anything else. Often the only recommendation they have is that they are in bad odor with some of the nobility on the other side, and professors of religion go to see these people. . . . But thank God the tide is turning. An actress came over here a short time ago and announced that she would make a tour of the states but the respectability of the land rose up against her and the engagements were canceled and she returned home a woman unfit for either continent.

Almost ten years later the Ash Grove Church of Christ lumped attendance at the theater with other heinous sins in the following edict signed by the elders of the church:

We believe that the Bible clearly teaches that the modern dance; card parties and card playing, either for amusement or for a wager or prize; theatre going; fornication; adultry; stealing; lying, perjury; the use, sale, and manufacture of alcoholic spirits as a beverage; betting; covetness, neglect of the worship of the house of the Lord; and all other disorderly conduct are hurtful and pernicious, both in themselves and in their consequences, and that those guilty of these sins should be subjected to discipline by the Elders of the congregation.[9]

Unrepentant trespassers of these proscriptions were promised that the fellowship of the church would be withdrawn from them.

The scorn in which professional plays and players were held was not for the most part transferred to their amateur counterparts. Local groups produced plays in nearly every community, usually to the plaudits of the press:

The Cassville [Barry County] Dramatists covered themselves with glory Thursday night in the presentation at the court house of "Down by the Sea." That they

had ample encouragement was testified by the packed
house, even the standing room being taken and many
applicants for tickets had to be refused. . . . The play
was given twice and total receipts from the two performances was about $70.[10]

Mrs. May Kennedy McCord recalled, however, that when
she was a girl of sixteen in Galena, in Stone County, the
family minister was upset with her title role in *The Gypsy's
Daughter.* He called at her home after opening night, she
said, and prayed over her, telling her parents that such a
role would cause her to become "worldly—may even lead
her to be an actress!" A direr fate would have been hard
for the Ozarker to imagine.

Many of the recreational activities of the Ozarkers
were closely connected with their way of life. A log rolling or a barn raising, in which groups of neighborhood
men and boys gathered for a day of exhausting physical
labor to help a neighbor construct a barn or a house, is
a prime example. "It was the collective nature of such
undertakings which contributed the recreational aspects,"
according to W. O. Cralle. "A group of people socially
starved welcomed this opportunity to meet with people
whom they had not seen for some time."[11] Other such
work-related collective activities included apple peelings,
cornhuskings, and quiltings. Recreations not so closely
connected to the business of making a living, such as picnics, play parties, square dances, and singings, also provided diversions for Ozarkers in search of amusement.

So most of the Ozarkers' entertainments were of their
own manufacture. In some the individuals who gathered
merely worked or played together (as at a cornhusking
or square dance) and took what entertainment value they
could from the activity itself and from the fellowship. But
a number of entertainments seemed to become almost theatrical events; that is, these entertainments contained the

17

theatrical elements—a performance, performers, an audience—usually associated with more formal dramatic production. Such entertainments seemed to have particular appeal to the Ozarkers, as evidenced by the amount of newspaper space and emphasis given to them and the enthusiasm and spontaneity with which such activities were recalled many years later in personal interviews.

As enjoyable as play parties, square dances, cornhuskings, and other self-entertainment might have been, they lacked the particular attraction of live performers appearing before audiences for the purpose of entertaining them. Ozarkers hungered for the peculiar excitement of this audience-performer relationship and often inserted it into entertainments that did not, by their nature, require this treatment. For instance, the function of an individual acting as Santa Claus at a Christmas Tree was to distribute the presents, while the pie-supper auctioneer's main job was to take bids on the sale items. Yet each of these functionaries was elevated to the position of performer and was expected to amuse and entertain the crowd while carrying out his other duties.

Ozark audiences also tended to stress the entertainment aspects of activities that were essentially nonentertainments by emphasizing their theatrical qualities. Religious services, for example, were rather universally regarded as entertainments, and the sight of worshipers caught up in an ecstasy of shouting or of a group of converts being baptized in a river was a spectacle to be enjoyed, while the preacher was also a performer whose message was perhaps of less importance than the manner in which it was delivered. Closing-of-school exercises became so highly entertainment-oriented that the teacher often felt compelled to present a program that would not necessarily demonstrate the pupil's knowledge but would compare in entertainment value with the program of the school in the next district. Finally, it is obvious that the

A large crowd witnessing a baptizing on Railey Creek, near Galena, Stone County. Photograph by O. W. Carter. Courtesy Mrs. Agnes Allen.

official purpose of a hanging was in no way to entertain anyone, but even the condemned man would occasionally become caught up in the holiday spirit and sing and talk to the huge crowd with all the abandon of a stage performer.

The desire of Ozarkers to be entertained caused them to congregate as *audiences* on many occasions, sometimes traveling for long distances and undergoing considerable discomforts to witness theatrical folk entertainments. Their appreciative response to the different kinds of performances encouraged the performers and helped assure the popularity of such entertainment. Many of the events allowed the audience to participate actively, as in group singing, speaking at closing-of-school exercises (or quizzing the scholars there), or exchanging quips with the auctioneer at a pie supper. Certain performers, such as champion spellers and debaters, ministers, and pretty girls (who might win the "beauty cake" at a pie supper), often gained followings that might best be described as fan clubs. These devoted groups actively supported their idols by whatever means were most appropriate, monetary contributions, moral backing, or rave reviews in the weekly papers. The audience-performer relationship was basic to theatrical folk entertainments, and whenever an Ozarks audience gathered for the purpose of entertainment, it could be assured of an interesting variety of performances and an equally interesting diversity of *performers*.

The Ozarkers' love of good talk is perhaps the reason why so many entertainments featured speakers of one kind or another. Preachers and political orators filled key roles at appropriate entertainments, and debates were held regularly by most literary societies throughout the Ozarks. Patrons rose to address the scholars at nearly every closing-of-school occasion, and condemned men about to be hanged used the gallows as a platform from which to address an absorbed crowd.

Individuals performed for the entertainment of audiences as pie-supper auctioneers, spellers, Santa Clauses, and kangaroo-court lawyers. Children and adults alike recited "pieces" or acted in dialogues and tableaux, and those with the necessary talent played musical instruments or sang.

Performances were given by groups formed to entertain by presenting band concerts, playing baseball, or putting on amateur plays. Audiences enjoyed also the spectacle provided by the emotional participants at a revival meeting or the sight of a queue of converts being immersed at a baptizing.

In lieu of a play as the *occasion* for the gathering of an audience, theatrical folk entertainments featured exhibitions of many kinds on almost any conceivable occasion. Some of these occasions had a significance or a purpose beyond entertainment. With others, the entertainment itself was the occasion. Box and pie suppers were conducted primarily for the purpose of raising money to benefit the school or for some other worthy cause. Religious gatherings were designed principally to help people live uprightly in this life and to give them an assist into the next world. Closing-of-school exhibitions had as their purpose displaying the knowledge of the scholars, while court weeks, hangings, and political speakings had their obvious purposes. Yet all of these events were particularly popular with audiences as entertainments.

Sometimes the entertainment was itself the occasion for the gathering of an audience, with no greater significance asked or needed. Literary societies, picnics, and band concerts belonged in this category, as did baseball games and community Christmas Trees. Whatever the occasion for an entertainment, it was usually of enough significance to assure that it would be well attended by an interested and enthusiastic audience.

The *settings* for Ozark entertainments did not often take

the form of specially decorated scenery. They did, however, provide a particularly appropriate background or environment in which the entertainment could be presented and thereby added much to the audience's enjoyment of the occasion. Settings were sometimes created by decoration. For closing-of-school programs, children's-day exercises, and community Christmas Trees, the school or church or courthouse was usually appropriately festooned with colorful decorations to honor the occasion. The setting for an Ozark folk entertainment was usually related to that entertainment in a peculiarly harmonious way. The nature of the entertainment determined the setting in which the event took place. The setting contributed to the total effect, subtly and unobtrusively, and was so closely allied to the entertainment that it is difficult to imagine the event in any other setting. A pie supper for instance, could theoretically be held in almost any location—a private home, outdoors, or on the stage of an opera house—but the association in the minds of most Ozarkers of the pie supper with the schoolhouse, where the supper was most commonly held, caused the atmosphere of the schoolhouse itself, brightened by the tempting mystery of gaily wrapped pies, to seem particularly "right" for that event.

The "right" setting for a minister's regular sermon was, of course, a church house, while the atmosphere of the courtroom was appropriate to the spectacle of the trial. The "right" setting for many entertainments was the outdoors: a baptismal service beside a clear running stream, a religious meeting under a canopy of limbs and leaves, or an open-air band concert beneath the stars.

A very important element of the setting was any element that seemed unusual or different to the Ozarkers —pageantry, color, noise, and crowds. Most rural Ozarkers led essentially lonely lives and welcomed any opportunity to gather with others. When the occasion for gathering was entertainment, anything in the total setting that made a strong sensory impression must be considered as a part of the setting of that entertainment. The total setting for the speech of the Fourth of July orator, for example, included not only the decorated platform or bandstand on which he stood but also the excitement, the cry of the lemonade vendor, the circle swings, the exploding firecrackers, the press of the crowd, and the remembered color and spectacle of the parade that had almost certainly preceded the speech. In a similar manner the settings of baseball games, hangings, political rallies, and court weeks were established for their Ozark audiences.

Theatrical folk entertainments—those entertainments that included performances by performers for an audience—were particularly pleasing to Ozark audiences because of the theatrical excitement surrounding this kind of activity. These entertainments included literaries, closing-of-school programs, religious gatherings, local dramatic productions, box and pie suppers, and picnics. It is these events that are described in the pages that follow.

"Literary Societies are all the rage."

—DOUGLAS COUNTY HERALD, NOVEMBER 3, 1892

Literaries

F RIDAY evening in the Ozarks was "literary" night. Some communities, it is true, had their weekly get-together on another day, but the phrase "Friday night literary" was commonly recognized as identifying the most common and popular entertainment of the period.

For many, getting there was half the fun of going. Patrons arrived at the school house by every available means of transport—lumber wagon, buggy, horseback, and of course, on foot. Grover Denny, of Texas County, would "take an old wagon and put three or four spring seats in it, and we'd all go in a bunch. We'd sing and just raise hell along the road. Nothin' meant by it. We just had a good time."

In the country there were not many buggies. The roads were simply too bad to make them practical. A correspondent in Howell County complained in print: "All along one of the largest and most travelled roads in the south part of the county the large loose stones or boulders are so numerous that it is dangerous to drive faster than a walk with a buggy or carriage and also very hard on a loaded wagon as well as dangerous."[1] The situation was no better in Webster County: a letter to the editor in the

This mule-drawn wagon may have been sturdy enough to negotiate the early-day roads in Taney County. Courtesy Douglas Mahnkey.

Marshfield Chronicle of September 1, 1898, commented, "On our present road you can neither talk or hear and if you have any eyes it takes them to watch the gulleys, stumps, and rocks to keep from getting your necks broken."

A young man might take his girl to the literary by providing the extra horse for her, or she might swing up behind him on his horse. The girl would seldom refuse to go because of transportation troubles. Claude Hibbard, of Douglas County, believed that "girls were different then in one respect. They would walk a hundred miles at that time if you led her by the hand and walked with her."

Going to a literary in the Ozark hills was sometimes complicated by meandering streams that must be crossed (sometimes two or three times) at treacherous fords where a slight miscalculation would plunge a horse or team into "swimming water." It was less dangerous but surely most embarrassing in Greene County when "a young gentleman while accompaning [*sic*] his best girl home from the literary made a misstep and fell in the branch and she followed."[2] The rugged countryside of Douglas County was the undoing of one young lady who was returning home from a literary meeting after dark when she lost the road and fell off a bluff.[3]

Oscar Morrill, of the Notch Post Office, in Stone County, told of being thrown onto a rocky road by the mule he was riding to an entertainment and tearing the seat of his trousers very badly. He repaired the damage by pinning the torn place together with large thorns from a nearby bush. He then remounted his mule and was on his way.

Whatever the difficulties, they were overcome in the search for entertainment and fellowship. Sam Miller, of Fordland, in Webster County, summed up the prevailing attitude: "I'd walk eight or ten miles to go to something, and when I got there, it wouldn't amount to anything. We were hungry for something to do."

One of the White River trails in Taney County over which
early Ozarkers traveled to literaries and other entertainments.
Courtesy Mrs. Jessie May Hackett.

Unlike box and pie suppers, the literary was seldom held for the purpose of raising money for any project. There was no admission fee, and the prime purpose was the entertainment of the audience and the education and enlightenment of both participants and audience. At the Shell Knob School, in Barry County, "There was an entertainment by our boys at night which caused some hearty laughs and people to forget the dry weather and hard times for an hour or two at least."[4] At Ash Grove School a different motive was given: "The young men of this place are trying to elevate themselves to a social political degree by coming out to the literary where they speak in a most distinguished manner."[5] Among the different kinds of literary programs were debates, kangaroo courts, spelling bees and ciphering matches; declamatory and dramatic programs, and special literary features.

DEBATES

Ozarkers loved their debates. Sam Miller said: "Why I've ridden ten miles to hear some debate. Some little old debate." Almost every literary, whatever else might be on the program, included a debate. Sometimes the debate was the entire program.

The interests of the Ozarkers, as reflected in the subjects they debated, were far from provincial. A great range of ideas occupied their interests and were explored through argumentation. Debate topics ranged from the contemporary political question ("Resolved: That absolute free trade would be for the best interests of the American people") to the universal, and often unresolvable, issue ("Resolved: That fire is more destructive than water"). Debates on issues of the latter variety were often humorous and most entertaining to the audience, and were therefore often reported in considerable detail by newspaper correspondents:

28

The cow vs. the sheep held the audience last Friday, verdict for the cow. Next Friday . . . the subject for debate is the dishrag vs. the broom and the fur will fly as all the distinguished orators have lined up and are consulting ancient and modern history for facts and figures. The writer is on the dishrag side as he has some sad memories of the use and abuse of the broom when wielded by the hands of an irate housewife.[6]

A correspondent wrote tongue in cheek: "The question for debate last Saturday night, Resolved, that life is not worth living, was decided in favor of the affirmative which has made us feel so bad that I wasn't able to eat but seven biscuits for breakfast."[7]

Upon noticing that the literary at the Scott schoolhouse, in Greene County, was to debate "Resolved: That woman is more attracting to the eye of man than monkeys," a correspondent with a sense both of fair play and of wit inquired: ". . . how many monkeys will be present at the aforesaid debate to bear witness just how attractive a woman is to a monkey's eye? You should tote fair with the monkeys, gentlemen."[8]

The Lilly School District, in Ozark County, became engrossed in the topic "Resolved: That the sun revolves around the earth." On February 6, 1896, the correspondent of the *Ozark County News* reported that "the opinion of the community is pretty well divided on this subject and both sides are well supported." At the third debate on the subject J. W. Curry made a "strong and well defined argument in favor of the affirmative which turned the tide of opinion to a remarkable degree." Decision went to the affirmative.

In 1906 residents of Ash Grove were hoping that an electric railroad would be built through the area. The train would stop at convenient crossroads for the accommodation of farmers who wanted to go to Springfield or

Ash Grove to do their marketing. Some Ash Grove merchants opposed the railroad, of course, fearing that their trade would be injured if the markets of Springfield, eighteen miles away, were made easily available. The town-country friction became apparent in such debates as "Resolved: That business men who oppose the building of Electric Railways and similar modern enterprises should be boycotted by the country." This particular topic was discussed at the Rock Prairie schoolhouse, in Greene County, and the correspondent commented: "This is a subject that has already been decided in the minds of many, but like most questions has two sides to it. We are in favor of a free, open and fair discussion of such matters to the end that all may become more alive to their importance."[9]

From the topics debated, the Ozarkers spent their time (1) considering historical, philosophical, ethical, and moral questions, and (2) debating current issues of international, national, or local import, with somewhat more emphasis on the former. The following list of debate topics was gleaned from the newspapers of the period:

1. Historical, Philosophical, Ethical, and Moral Questions

Whiskey versus war:
RESOLVED: That whiskey destroys more lives than war.
That more misery has been caused by intemperance than warfare.
That war has been more destructive to the human family than intemperance.
That intoxicating drink has caused more pain, sorrow, and woe than all the wars combined.

Love:
RESOLVED: That love is a greater incentive than the fear of future punishment to the action of man.
That love is a stronger incentive to human action than money.

RESOLVED: That a man will go further for love than for money.

The Indian:

RESOLVED: That the Indian received more cruel treatment from the white man than the Negro.

That the Indian has a better right to North America than the white man.

Education:

RESOLVED: That war has done more for education than books.

That more information is obtained by traveling than by reading.

That education is not the greatest source of happiness.

That education increases crime.

Miscellaneous:

RESOLVED: That fire is more destructive than water.

That married life is happier than single life.

That slander is more injurious than flattery.

That metals are more useful than animals.

That country life is more desirable than city life.

That the sun revolves around the earth.

Personalities:

RESOLVED: That Benjamin Franklin deserved as much praise in producing the Independence of America as Washington did.

That Lafayette deserves more honor from the American people than Washington.

That Columbus deserved more credit for discovering America than did Washington for defending it.

That the position of Lincoln as President was harder to fill than Washington's.

That President Jefferson's administration was more beneficial than President Lincoln.

That Winfield Scott was a greater general than Zachary Taylor.

That Grant was a better general than Lee.

That Pilot [sic] was a meaner man than Judas.

That Judas Iscariot was once a Christian.

Printing press:

RESOLVED: That the printing press had done more for the advancement of civilization than the railroads.

That the printing press is a greater benefit than the steam engine.

Man's nature:

RESOLVED: That a man's life is just what he makes it.

That all men are cowards.

That surrounding circumstances make a man's character.

That man is the architect of his own fortune.

That every act of man is selfish.

Woman:

RESOLVED: That woman has more influence on man than money.

That woman is equal mentally to man.

That women have done more to Christianize the world than men.

That woman is more attracting to the eye of man than monkeys.

Religion and morals:

RESOLVED: That conscience is a true moral guide.

That nature proves the existence of God more conclusively than does the Bible.

That the world is growing morally worse.

That professors are more detrimental to Christianity than sinners.

That a Roman Catholic can be a good citizen of the United States.

That a Christian cannot go to war.

That life is not worth living.

That pride and ambition are a greater curse to the human race than ignorance and superstition.

Present conditions and prospects:

RESOLVED: That opportunities for young men are greater than they ever were.

RESOLVED: That the United States has reached the zenith and is now on the decline.

That the world has reached its zenith and is now on the decline.

The farmer:

RESOLVED: That menial labor is more important to the agriculturist than manual labor.

That the rural delivery was of more benefit to the farmer than the telephone.

Discussion questions:

Do the signs of the times indicate the downfall of the Republic?
Can a big man ache harder than a little one?
Was Andrew Johnson impeached while President of the United States or not?
In what century are we living [in 1900]?
Which is more powerful, the sun or the wind?
Which is the cause of more evil, women or money?
Which is the more useful to mankind, the broom or the dishrag?
Which is more attractive to the eye, art or nature?
Which is more useful, the sheep or the cow?
Which is more useful, wood or coal?

2. Current Issues

International:

RESOLVED: That the United States should annex the conquered territory of the late war with Spain.

That the war in the Philippine Islands is just and ought to be pushed.

That the President's policy in the Philippine Islands ought to be endorsed by every true American.

That the newly acquired possessions are beneficial to the United States.

That the Boxers are justifiable in opposing foreign invasion.

That imperialism if adopted will prove to be a great danger and a constant menace to the peace and stability of the United States and that every true

RESOLVED: American should take a firm stand against the policy of forceful annexation and imperialism.

That the United States should recognize the Cubans as belligerents.

That absolute free trade would be for the best interests of the American people.

That Japan's objections to the United States' annexing Hawaii are as well founded as the United States' objections to England furthering her territorial possessions in Venezuela.

That the United States should acknowledge the independence of Cuba.

That Cuba should be annexed to the United States.

National:

RESOLVED: That the World's Fair at St. Louis will be detrimental to the United States.

That the Negro should be colonized in the Philippine Islands.

That the construction of the Panama Canal will cost more than we will ever receive from it.

That women should have the same religious and political liberties as men.

That patronizing mail order houses is detrimental to the best interests of the country.

That life insurance companies are a detriment to a country.

That railroads are detrimental to a country.

That the income tax law is unjust.

That corporations are a menace to republican institutions.

That the Indian should be made to support himself.

That the manufacture and sale of intoxicating liquors should be abolished.

That women should have the right to vote.

That foreign immigration should be prohibited.

That the telephone and telegraph should be owned and operated by the government.

That the United States Senators should be elected by popular vote.

34

RESOLVED: That all money should be issued by the United States Government direct to the people without the intervention of any banking or other corporation.

That the United States should have the free and unlimited coinage of silver at the ratio of 16 to 1.

That modern inventions have been beneficial to the laboring man.

That the present wage slavery is worse oppression to the laboring class than chattel slavery was before the Civil War.

That bachelors should be taxed to support widows and orphans.

That women should be excluded from the commercial world.

Social:

RESOLVED: That to restrain sheep and hogs from running at large would be of benefit to Roaring River township.

That all persons between the ages of six and twenty years should be compelled to attend school four months each year.

That orchards are financially a detriment to the country.

That the public school system as at present conducted is a nuisance and a fraud and should be abolished.

That Missouri should have a compulsory education law.

That the coeducation of the sexes is advisable.

That the Bible should be taught in public schools.

That business men who oppose the building of Electric Railways and similar modern enterprises should be boycotted by the country.

Political questions were debated along sharply partisan lines, and decisions were rendered in the same manner. Brushy Knob Society debated the free-silver issue in 1898: "The affirmative was represented by five Democrats, and the negative by five Republicans. The Republicans tied the silver men up in knots and had them beat from the start. . . . The vote resulted in 30 to 8 in favor of sound money. Any old thing can best the silver question now, even little school boys."[10] Except on political questions, however, a high-minded attitude of free and open discussion and fairness seems to have prevailed. Decisions of judges were usually not seriously contested, and often decisions were left to a vote of the entire audience. Seldom was there such dictatorial control as in the society at Exeter (Barry County): "Friday a debating society was organized at Beasley Hall. J. O. F. Beasley was elected president. No by-laws were drafted. All questions are to be left to the president who also decides all debates."[11]

Debate teams were sometimes made up of schoolchildren, but more commonly they were formed of adult members of the community who, in Vance Randolph's words, "could think up a powerful good speech in a whole week of walkin' down the corn rows."[12] At some debates two champions chose their colleagues until there were about four on each side. These leaders made the principal speeches, allowing the others to say a few words whenever they could think of a point to make.[13] In some societies the topic was decided on the night of the debate, and discussion was strictly impromptu.

When a district developed one or more favorite debaters who could be expected to uphold the honor of the society, a challenge was sent to a neighboring community for a meeting of champions. The challenge was sometimes issued through the columns of the local newspaper, as was this one from Seligman: "The question, Resolved: That each man's life is what he makes it, has been debated at almost every school house in this township. Two men, Roller and Murray from Seligman, are ready to meet in debate any other two debaters who may favor the affirmative."[14]

It was in a newspaper that a squabble occurred over

a debate that never took place. Through the newspaper the Corsicana (Barry County) Society accused the Blackstone Society of "failing to ante" to a challenge to debate. The Blackstone representative replied heatedly:

> I say we did accept the challenge and insisted that they comply with the rules governing all debating societies and come to our school house. This they refused to do. . . . They have a literary paper which stated that Central Society (Blackstone) was dead. Now boys, if you will only come to parlimentary rules and meet us you will be made to believe that you have met a live corpse.[15]

The style of debate varied widely. One man who attended a debate at Scott schoolhouse, in Greene County, was impressed by the "minds and intellects" in the society and wished for equal strength in the halls of the legislature.[16] Jason Roy told of an elderly man in Douglas County who in his debates tried to use big words that were sometimes far from correct in meaning: "We learned to recognize the sincerity of that man and we never at any time laughed at him because he was sincere and tried to help out. He thought it was quite a brightness to utter big words."

One of Vance Randolph's characters in *Ozark Mountain Folks* discusses debating tactics:

> The main thing in debatin' is to get a chanct to speak first and then jump in and mention ever' pint on the other side sayin' how them is pretty good arguments but ever last one of 'em has got a hole in it some whar and when t'other side gets up they cain't do nothin' only kinder foller long after what you done said cause there ain't nothin' new fer them to fetch out and the jedges mostly thinks well probable he wouldn't ever never figgered out all them things if your side hadn't said 'em first and peers like he ain't makin' no great go out of it no how.[17]

37

Uncle Joe Cranfield remembered a "pint" made by a debater that did him no good:

> Well we had . . . a debate about the darkey, how he was treated. Well one boy got up and he was just amakin' this as he went—he said, "Them darkies didn't get nothin' to eat down there when they was slaves only rolled oats and they hadn't never been shelled— they jest got trash and all together." Well that purty near broke it up and he lost over that.

Mrs. Eva Dunlap recalled with pride her victory on the question "Which is more useful, coal or wood?" Supporting the wood side, she crushed her opponent by pointing out that trees grew from the ground and dropped their leaves, which in turn were compressed underground to form coal. Her triumphant conclusion, "So if it hadn't been for trees there wouldn't have been any coal at all," drew the applause of the audience and the decision of the judges.

She told also about the elderly man who, while debating, became offended at the personal remarks of a member of the opposition, picked up the coal-oil lantern that he had brought to the schoolhouse and stalked angrily out of the building. Since the lantern had provided the only light for the debate, the meeting adjourned sine die.

KANGAROO COURTS

Lawrenceburg has a moot tribunal of justice known by the uphonious of "Kangaroo Court." It takes cognizance of all manner of real or imaginary offenses "against the peace and dignity of the town." The witnesses are sworn not to tell the truth or anything resembling it. It is a remarkable fact that the poor culprit whose case comes up before this bar has never been known to escape conviction.[18]

The kangaroo court as a feature of the literary program was not universal through the Ozarks, but in the areas where it was practiced it rivaled the debate in interest and popularity. Procedures in kangaroo courts ranged widely, but the description of the Lawrenceburg court given above is an indication of the manner in which most of them were conducted.

The person selected as the defendant was hauled before the court and charged with an infraction of an imaginary law. The indictment was often farfetched but was usually in some manner based upon fact or the actual conduct of the defendant, if only as the antithesis of real facts or behavior. A teetotaler, for instance, might be charged with drunkenness, or the president of the literary society with ignorance. Making big eyes at the girls, sleeping during church or a literary, and sparking during literary hours were some of the charges recalled by Tom Rideout and Ambrose Renfro, of Howell County. Uncle Joe Cranfield recalled kangaroo courts:

> We'd have some of the awfullest times there ever was. This old man Harrison Hall he put in a case in the kangaroo court about this neighbor that he'd stole Aunt Bess Garth's old sow. Well they was just ahavin' it around and around. First thing Harrison Hall knowed they slipped in a witness that swore that Uncle Harrison Hall stole Aunt Bess Garth's old sow and took her off, they seen him agoin' with her. That just like to a-ruint 'em out at the Kangaroo court.

The punishment for a conviction (and the defendant seldom escaped being found guilty) was usually a fine, a nickel to a dollar, which was used to buy coal oil for the school lamps or to pay someone to clean up the schoolhouse. At an Ash Grove School literary, "One man was fined fifty cents for not marrying and another five cents for letting his calves drink up another man's sugar water."[19]

E. H. Vandiver, of Brandsville, in Howell County, indicated that the proceedings of the kangaroo court were
good-natured and friendly: "Some stranger would come
in here and we wouldn't let him be too much of a stranger.
Some of us would go out and arrest him and haul him in
here and pour it on him."

Most of the courts were rather loosely organized, with
a pickup judge and jury and whatever other officers of the
court were deemed necessary. The position of attorney,
either prosecuting or defense, was the most important,
and usually only a few men of the community—the "best
gabbers," as Tom Rideout characterized them—were admitted to practice.

Joe McKinley, who began teaching in Wright County
in 1896, described a very elaborate and well-organized
court. Lawyers were willing to come out from town, he
said, to serve as judge and counsel, and the cases were
tried in accordance with actual legal procedure. Kangaroo
court was held only once a month to allow adequate time
for preparation, with ciphering matches and spelling bees
filling the time in between. But neither of those events
held the intense interest of the community that the proceedings of the court held. The participants took their
work seriously, but no more than did the people of the
community who were observers.

When McKinley was about eighteen years old, he sued
another young man, who lived in town, for the alienation
of his girl's affections:

> It was a friendly affair, you know. However, I had
> kept company with this girl just along. We just agree-
> ably each one went our way, and she immediately
> began going with this boy from the town. Well you'd
> be surprised how keen that became to those people
> out in my community, five miles from the town where
> this boy lived. Boy! They went into it just like it was
> real. I had been imposed upon! . . . I won my case, and

he was supposed to pay me a penny a year. I relieved him of all responsibility since he asked for a receipt each time.

The kangaroo court, as practiced in McKinley's school, may have instructed the participants and observers in the facts and procedures of the law, but it is doubtful that major benefits accrued to the audiences. The courts' popularity was due to the combining of the inherent drama of the courtroom situation, the Ozarkers' love of speechmaking, and their delight in tall tales, expressed in the courts in the fantastic charges and proofs brought by both sides. It was a happy combination, and one that provided many an hour of respite from the drudgery and toil of everyday existence.

SPELLING BEES AND CIPHERING MATCHES

The ability to spell correctly was regarded as a most desirable educational achievement. The skill was displayed publicly to the admiration of a community's citizens, most of whom perhaps not as far advanced in academic achievements. The spelling bee, with words pronounced from the *Blue-backed Speller,* became a popular feature of community entertainment and found acceptance as part of a literary program an entertainment in itself.

School patrons were concerned about the level of literacy of their children. "A generation of stammerers and poor spellers," wrote a rural schoolteacher in 1894, "bears witness to the aimless and lifeless ways of teaching spelling (or rather of not teaching it) which have been so common for about fifteen years."[20] The ability to spell orally seems to have been taken as prima facie evidence of an educated person. Thus the popularity of the spelling contest lay not only in the excitement of the elimination contest but also in the tangible evidence of the educational benefits of the school.

The best spellers were usually selected by the teacher, and they alternately chose others to be on their teams. Sometimes the schoolchildren were the contestants, but the patrons often participated as well. The object, of course, was to "spell down" the other side. If a contestant could not spell a particular word, he sat down, and the word was passed to the other side. This process of elimination was continued with words progressively more difficult, until the contenders had been eliminated and only the champion remained standing. Words like "kaleidoscope," "cuisine," and "caoutchouc" were read out and spelled to the great admiration of the audience. In some bees it was necessary not only to spell the word correctly but to re-pronounce it, syllable by syllable, in a tongue-twisting phonic drill. "Extemporaneous," for instance, would be correctly rendered thus:

Pronounce:	"Extemporaneous."
First syllable:	"E-X, ĕks."
Second syllable:	"T-E-M, tĕm. Ĕks-tĕm."
Third syllable:	"P-O, pō. Ĕks-tĕm-pō."
Fourth syllable:	"R-A, rā. Ĕks-tĕm-pō-rā."
Fifth syllable:	"N-E, nē. Ĕks-tĕm-pō-rā-nė."
Sixth syllable:	"O-U-S, ŭs. Ĕks-tĕm-pō-rā-nē-ŭs."

Those individuals who could perform such verbal calisthenics took considerable pride and enjoyed the admiration of their fellows.

Like the debaters, spelling "stars" or champions emerged and traveled from school to school, wherever a spelling contest was held, followed by an admiring coterie. It was a point of great pride to be recognized as a good speller and to be chosen early in the process of choosing sides. The important thing, of course, was the satisfaction of winning, of being the last contestant standing, but at some bees there were tangible rewards. A quilt was the prize for the best speller at the Methodist Episcopal church in

Houston (Texas County),[21] and at Ash Grove the winner
of a match received a book entitled *Representative Men*,
while each of the two spellers who went down first received a consolation prize, a Japanese doll baby.[22] This
match, reported by the correspondent in a breathless play
by play, describes the typical spelling program:

> The old fashioned spelling match in Chandler's
> Hall last Friday evening was well attended. After a
> song, "Old Folks at Home," Ote Weir and Ed Barbee
> were put up for class leaders. They chose sides until
> the audience was thinned out. Mrs. Lula Silver gave
> out the words from an old blue backed speller. Clint
> Nicholson on Ote's side was the first to go down. W. T.
> Chandler was the first on Ed's side to demand a divi-
> sion of the honors of the evening with Nicholson.
> Chandler insists he was in dead earnest but the crowd
> still believes he went down for the fun of it. . . . It
> was finally narrowed down to Mrs. Doyle and Henry
> Swindler. Directly Henry went down on "lethargy"
> and Mrs. Doyle was declared the winner. . . . After
> a song by Ed Barbee and some light refreshments the
> meeting was adjourned.[23]

Spelling bees were almost universal in the Ozarks,
though they were given more emphasis in some areas than
they were in others. They were always popular entertainments, and Ozarkers attending them usually went home
happy, as did the Ash Grove patron who reported that,
after a particularly interesting spelling, "Mush and milk
was served and everybody had a good time."[24]

The ciphering match was conducted along much the
same lines as the spelling bee but lacked its widespread
popularity. For one reason, the ability to figure, while
regarded respectfully as one of the triumvirate of R's, was
not held to represent "learning" to the degree of the ability
to spell. Too, the cold, logical symbols of mathematics
lacked the magic warmth of language. In dramatic appeal

the command "Divide six thousand, eight hundred fifty-three by one hundred twenty seven" and the resulting click of chalk upon blackboard or slate pencil upon slate could not compare with the satisfaction of the euphonious reply to a request to spell the word "ratiocination" or "ichthyology." In the spelling bee the contestant's strained concentration, the suspense of waiting for a response — in other words, the observation of the human mind in competition, was thoroughly absorbing.

The ciphering match was more popular and the interest more intense when the competition was between two schools. The patrons of the host school turned out in force, and the visiting champions brought a goodly representation from their community. A popular procedure for a ciphering match pitted the star of the host school against a star selected by the visitors. The victorious contestant in the first problem stood against the second challenger, and so on. A contestant ciphered as long as he could stand. It was not unheard of for a single champion to stand down an entire team of challengers. Another procedure was described by Emmett Yoeman (Douglas County): ". . . they would bring up two at a time, and they would usually draw for their choice then, whether it would be addition, subtraction, multiplication — usually it would be addition. Then another pair would come up, and it was kept track of on a point basis."

The contestants worked the problems on the blackboard, if the schoolhouse had one, so that all could see. Speed was essential of course, but form was no less important. The cipherer not only must arrive at the correct answer first but also must show all his computations in correct arrangement, including all carrying figures. McKinley told about a girl who was "hard to beat." She performed long subtraction problems in her head, wrote down the entire remainder from left to right, and then perfected the form of her problem by writing down the subtrahend

44

and the minuend, crossing out figures of the minuend, and jotting in the carry numbers. One can imagine the enthusiasm with which ciphering fans witnessed the performance of such a virtuoso.

DECLAMATORY AND DRAMATIC PROGRAMS

Phelps school, in Lawrence County, presented a typical "exhibition" one spring night in 1887. Patrons and visitors were treated to a series of recitations, three short plays, and a song. Some of the recitations were "The Mule Stood on the Steamboat Deck," "Dr. Puffstuff," and "Munnetunkee Chief," while the favorite farce of the evening was "Married by the New Justice." The "serio-comic song, Mary Has Run Away Wid a Coon," was another highlight.[25]

A debate was the feature attraction at Cassville (Barry County) in 1895, but the correspondent also mentioned with favor the declamations given by two young ladies, and was so impressed by one, "The Devil's Visit to One of His Servants, A Saloon Keeper," that he gave a résumé of it: "He was badly frightened and thought his time had come to stop breathing but his majesty told him that he was filling hell with drunkards and he must remain at his present business many years to come. That he was too valuable a servant to lose."[26]

In Ash Grove, in 1893, the "literary society free entertainment" featured

the Wild and Reckless Rascal Pat, a drama very humorous, two operatic quartettes one of which is We Will Never Mortgage the Farm, the other a laughable quartette entitled The Fortune Teller full of predictions and culminations. The fan drill by twelve young ladies, the shorthand medal contest, the best report of a speech to be delivered by Professor Bloomer will win the medal.[27]

45

These examples give a fair indication of the variety of events featured in Ozark entertainments. They were usually presented by the schoolchildren, the adult members of the community serving as audience. Such a program resembled closely, and was in many instances preparation for, the closing-of-school program, which was one of the main events on the Ozark entertainment calendar. Closing-of-school entertainments are described in detail in a later chapter.

SPECIAL LITERARY FEATURES

One event of a dramatic nature of about 1895, connected with a literary at Pleasant Hope School, in Greene County, deserves special mention because of the detail in which it has been reported, as well as the insight it gives into the community character of the literary, and the Ozarker's love of roasting a friend. At the literary a play was presented that grew out of an event in which Mrs. Eva Dunlap was involved. A bachelor neighbor, Press Rector, was absent from his three-room house one day when his older brother Mack dropped by to see him. Finding no one at home, Mack, for a joke, put a chair in Press's bed, pulled the quilt over it, and left.

Press soon appeared at Eva's house and excitedly announced that when he got home he found a tramp in his bed. Eva's father and uncle prepared to accompany him back to his house to investigate. This is the story, as Mrs. Dunlap told it:

> As they passed the woodpile, Papa he picked up an ax and Aunt Teen came out and said, "Oh Joe, don't go! Don't go! You know I had a dream last night, and you know that awful dream I had and don't go!"
> Uncle Joe went along just the same and Press went on ahead. And I got up and followed with 'em. Aunt Teen and I both went. We got on up there and

46

The pupils and teacher (behind the pupil fourth from the left) of Evening Shade School, in Ozark County, line up for the yearly school photograph. Courtesy State Historical Society of Missouri, Columbia.

there were three straight rooms in a row. Well we
went into the one fartherest on the east and had to go
through this middle room, and this man was in the
bed in the front, back in that corner. And there was
a fireplace in that room back there, and Press's revolver lay on that mantle, and he had to pass right
by that bed like this, to get his gun and come back
to that middle room, and there stood Uncle Joe, and
he was a-sayin', "Gettup my friend! Gettup my friend!"
And there wasn't any move made, and Press got around
here and got his revolver and just pulled the trigger,
you know, and he shot right through the bed and mat-
tress. He didn't shoot to hit the man, he just thought
he'd scare him. And still he didn't move, and he went
in and jerked the cover off, and he said, "It's a cheer."

Well, Uncle Joe and Papa just screamed and
laughed, and they went out of there and Aunt Teen
was still asayin', "Don't go! Don't go in there!"

Then after that was when we had the play up at
school, and we acted it out, you know. I acted my
part and Clara filled in for Aunt Teen. She begged
Joe not to go. . . . I don't know who took Uncle Joe's
place, or who took Press's place, but Press sat back
in the audience with his head down on his desk like
this and he never raised his head until that was over,
and I'll tell you, that crowd that knew anything about
it just screamed. Well, they all laughed, but a lot of
them didn't know what it was about. We had a house-
ful of people up there.

Local events were more commonly reported, exaggerated, or manufactured in the "literary paper," a jour-
nalistic effort read at the program and received with great
approval by the audience. Tom Rideout said that the editor of their paper "made up jokes on other people and
read it off to the public." The New Site (Greene County)
paper, the *Headlight,* was characterized as "one of the spici-
est little papers in the land."[28] At Pleasant Hope School

a correspondent once noted that for the first time since the society was started there was no literary paper. She explained, "The reason why there was no paper was because our editors took part in the entertainment at Bois D'Arc [Greene County] Saturday night for which they had to spend their time practicing last week."[29]

A very special feature of a literary program, which held great delight for the audience, was the tableau, illuminated by the colored flame of burning powders. Mrs. May Kennedy McCord said that it was a great honor to "have the powders burned on you," eclipsing even the opportunity to sing a solo.

The illuminated tableau was not presented very often because the powder cost too much—fifty cents for enough for two or three burnings. Its position was inevitably last on the program, for the burning powder emitted clouds of thick smoke that filled the schoolhouse and could be dissipated only by throwing open all the windows and waiting until the air cleared.

Mrs. McCord described a tableau that took place at the turn of the century at her childhood school in Galena. After the main part of the literary was completed, men stationed around the schoolhouse extinguished the oil lights. The audience sat in darkness for a few moments, awaiting the colorful spectacle they knew was in store. Then someone touched a match to the handful of powder in the school coal shovel, which lay close to the edge of the platform. The audience had not been aware that a temporary curtain had been drawn during the brief blackout, and now the pink flare disclosed a girl clothed in a white flowing robe, her long hair rippling to her waist, clinging to a large cross mounted against the blackboard at the rear of the building. The green material engulfing the feet of the girl was recognized as representing the sea. An offstage voice declaimed the words of the hymn "Simply to Thy Cross I Cling," while the audience gasped

49

in wonderment. For thirty seconds the powder burned, then sputtered and went out. Twice more the powder flared, first revealing a scene called "The Proposal"—a garden bench, a lovely girl in a cartwheel hat, and a young man kneeling at her feet. The children in the audience snickered, and the young man in the tableau looked pained. The final tableau was a solemn one, showing a thoughtful Lincoln sitting at a desk, writing. The audience was respectful in the presence of this representation of the great Republican, but the smoke was causing them to cough and sputter. As the soft light died for the last time, some men relit the lamps while others opened the windows. The literary was over.

An endless variety of specialties spiced Ozark literary programs. Inevitably, blacks were the butt of humor. Mrs. Dunlap told of "nigger huskin's," an entertainment in which a group of white couples sat around a pile of corn on the rostrum with black cotton stockings over their heads and exaggerated red lips painted on and, while shucking ears of corn, told "blackout-type" jokes and stories about others of the community.

Some literaries had elaborate musical entertainment, as at Phelps (Lawrence County), where "the string band led by George Miller sends forth sweet strains of music to the delight and satisfaction of all present,"[30] and at Granger City, where the "Granger City choir gave a song which was enjoyed by all."[31] In Howell County two men furnished music played on a cornstalk fiddle and a gourd banjo. Their reception was not lessened by the fact that the only tune they could play was "Isaac and Joshua."[32] Charlie Weaver, of the Weaver Brothers and Elviry troupe, began playing with his brother Leon at Ozark literaries when he was ten years old.

Soon after the turn of the century a Graphophone craze struck the Ozarks, and an entertainment was hardly considered complete without one of the "talking machines."

A typical report appeared in the *Houston Herald* of December 5, 1901:

> The graphophone last Friday night was attended by a good sized crowd and proved an interesting entertainment. Mr. Blankenship has one of the finest instruments made and his variety of records covering a program of songs, solos, and orchestra and band selections, humorous and oratorical addresses is such that all may be pleased.

A successful program at one literary was likely to be copied and repeated by other groups. The "Peak (or Peaked) Sisters" visited literaries around the area, wearing conical paper hats two feet tall, carrying bandboxes containing combs and tissue paper and other strange musical instruments, and entertaining the audience by acting "green" and with recitations and song. In Houston they were known as Sister Keziah, Sister Mariah, Sister Sophia, and Sister Betsey (who was deaf and dumb). The reporter stated enigmatically that "the attire of the sisters was unique."[33] At Mount Vernon (Lawrence County) they became Vindy, Windy, Zantippy, and Zaloony,[34] and at Galena, where they were nameless, their number increased to seven, and they so convulsed their hosts that "the entire audience was in pain from excessive laughter."[35]

Another peripatetic favorite was "Mrs. Jarley" with her famous "Wax Works," which she set up in the local schoolhouse. A curtain was pulled revealing a series of "wax figures" in costume and in poses indicative of the characters they portrayed. Mrs. Jarley went from "statue" to "statue," winding an imaginary spring that activated each figure for a brief moment of animation. Mrs. Dunlap described some of the characters:

"Jesse James Dusting a Picture" (his moment of life lasted just long enough for him to be shot).

"Indian Scalping Girl" (Mrs. Dunlap recalled no significant action other than that the girl did get scalped).

"Mermaid Lying on Sand" (a girl wearing a fishtail lay on a camouflaged bench and sang "Down at the Bottom of the Sea"; this act stopped the show).

"Giant Eating Bread and Milk" (a boy stood on a chair wearing a skirt that reached to the floor; when he was wound up, he ate bread and milk from a bowl).

"Maniac" (when animated, she pulled her long hair and screamed).

"Boy on Burning Deck" (his moment of truth occurred as he recited: "The boy stood on the burning deck / Eating peanuts by the peck. / Peanuts! Peanuts!").

Other characters were added according to the desires and imagination of the producing group.

Finally it should be noted that in the absence of debate, kangaroo court, spelling bee, or tableau, Ozarkers could still enjoy their get-togethers. A party from Forsyth journeyed to Taneyville (both in Taney County) to attend a literary. When the literary failed to materialize, they, undaunted, "all began to sing hymns and had a religiously good time."[36]

It can be safely stated that nothing is more attractive and interesting to the largest number of people than the exercises following the end of the session of an institute of learning.

— LAWRENCE CHIEFTAIN, JUNE 16, 1892

Closing-of-School Programs

THE young man with good recommendations from the Springfield Normal School who placed his bid to teach at the Mountain School in Lawrence County for the sum of $28.50 a month[1] might well have saved himself the bother, for the board had previously announced their disinclination to pay any such extravagant salary. The directors, "taking into consideration the great number of teachers in the field, the low price of farm products, scarcity of money and hard times generally," decided not to pay more than $20.00 a month, and warned those who placed a higher value on their services not to apply.[2] Besides the expensive applicant noted above, they received another bid for $25.00, but they had their choice of three teachers at the stipulated rate of $20.00. "The scholars are not advanced, are very backward," noted the correspondent to the local paper. "Then what is the use and where is the justice in employing a teacher at some fancy price?"[3] The patron of the Taney County school whose letter to the *Taney County Republican* appeared on September 28, 1899, agreed with this attitude. "We do not believe it is the high priced teacher that is the best qualified," he wrote, and praised the local schoolmaster, who was giving "general satisfaction" at $18.00 a month.

Many rural districts preferred to hire a local boy or girl who had completed the eighth grade and had received a teaching certificate either by completing a term at the "summer normal," or by passing an examination given by the county superintendent of schools. Such a person could afford to work more cheaply than an outsider, because he or she could live at home and need not pay room and board. The only pedagogical skills required were those that would enable the teacher to impart the fundamentals of reading, writing, and ciphering; maintain a modicum of discipline; and act as the producer of an entertainment at term's end.

The months of February, March, and April brought the closing of most of the six-month rural schools in the Ozarks, though some closed as early as Christmas week, while others continued until May and even June. On closing day the entire neighborhood went to the schoolhouse for "end of school" was celebrated with an entertainment event not only for those who had children enrolled in the school but for all members of the community.

Besides its high entertainment-quotient ranking, the end-of-school program had another important function: it was then that the merit of the just-completed term, and of the teacher, was evaluated. A schoolteacher who had conducted school with a minimum of disciplinary troubles and had avoided open conflict with the more influential patrons was tentatively adjudged to have taught "a most successful term." But the real moment of truth arrived at the close of school, when "everyone gathered to judge the manner in which the school had been conducted."[4] Emmett Yoeman, who taught in Douglas County said, "This program definitely had a bearing on whether the teacher was invited back the next year." In Barry County in 1895 a teacher was chosen by a unanimous vote of the school as

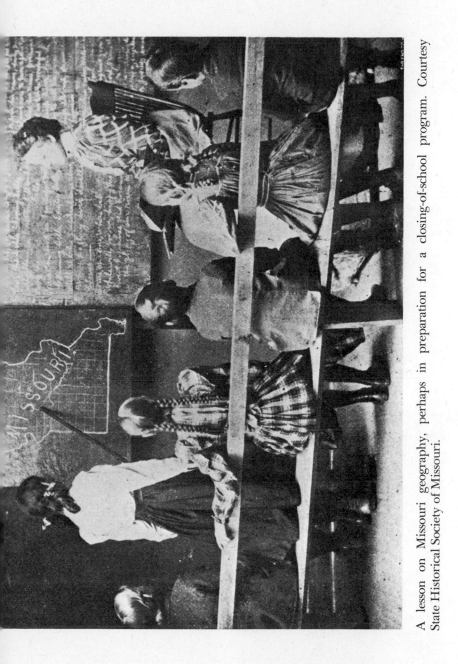

A lesson on Missouri geography, perhaps in preparation for a closing-of-school program. Courtesy State Historical Society of Missouri.

his own successor for the coming year, *following* a highly acclaimed closing program.[5] The program at Gobbler's Knob (Taney County) in 1896 "gave entire satisfaction and the patrons and children seem anxious to secure the teacher's services again for the coming year."[6]

The teacher found it a decided advantage to produce for the community a program that would compare favorably not only with those of his or her predecessors but also with those of surrounding districts as well, for patrons attended those exercises as well and made comparisons. Yoeman put it thus:

> [The closing program] didn't actually represent the work done at the school, but as far as the people was concerned, it was a comparative proposition. If a fellow showed up real well they'd say, "He's doin' a good job over there," and if he didn't, "Something wrong over there at Walnut Grove. He didn't do much." So each one did his darnedest to put on a good program, one that would entertain the people.

The last day was often divided into two parts, the first part being a display of the scholars' abilities in certain school subjects and the second part a literary-type entertainment. The forenoon exercises may have been an approximation of regular classwork, final oral examinations, or, more commonly, reviews in such subjects as arithmetic, geography, physiology, and language. The patrons of one school "viewed the drawings that were elaborately placed upon the blackboard in colors"; examined the term's tests, which were bound together; and listened to a drill on "'Memory Gems,' . . . culled from the ablest authors and sages of the past."[7] A patron of the New Site Community boasted of having heard "the best class in arithmetic we ever heard recite,"[8] and at the Phelps school, in Lawrence County, the patrons and visitors had the privilege of reviewing the classes themselves.[9]

The teacher who did not allow some time during the day for remarks by the patrons of the district did not care much for the traditions of the occasion, or for his position of respect in the community. Guy Howard, at the end of his first term of teaching in an Ozark school, made sure to observe this amenity: "It would have been a grave error to have omitted this closing day courtesy to the patrons. In the hill country one must be very sure that he has a substitute that will meet with ready approval before he deviates in the slightest way from the usual way of doing things."[10]

Newspaper reports of close of school often mentioned speeches by the patrons, as at Big Creek, in Texas County, where they "expressed themselves as being well pleased with the school and gave the pupils good advice and encouragement."[11] It was not uncommon for the visitors to recite "good pieces" for the entertainment of teacher and pupils. At Center School, in Greene County, Hugh West "began by apologizing for his lack of oratory, but before he got through a thirty minute address he showed that he is an orator of no mean ability."[12]

When the morning exercises were over and dinnertime (noon) arrived, the "well-filled baskets" that the patrons had brought with them were placed on "tables which groaned under the weight." This part of the program was enjoyed by all: "The most bashful could not resist the temptation of that bountiful and excellent dinner, and the students seemed to forget the physiological rules that had been taught them concerning eating too much just because it tastes good."[13]

So firmly entrenched was the format of the end-of-school entertainment that few teachers risked tampering with it. The classwork review might be eliminated, but the entertainment, never. A daring teacher at Ash Grove gave notice that his exercises would consist of a "general oral review and examination" on each of three days. He added:

57

"I believe class work is worth more to pupils than long drills for exhibits at the close of the session. Therefore I have kept up our regular daily program to the close." But even this pioneer capitulated to the extent of having a program, though he stressed that "the entertainment is not expected to be anything more than the recital of a few selections interspersed with music upon which but little drill has been given."[14] As an educator he was no doubt concerned about the great amount of time (sometimes as much as six weeks of the six-month term) normally given to the preparation of the entertainment program.

For those who attended a typical country-school closing the program consisted of recitations, dialogues or playlets, declamations, and music. Sometimes the programs were long. The invitation to the Phelps school closing in 1893 urged patrons to "bring your baskets supplied and remain for exhibition at night," and one hopes that the advice was heeded, for sustenance was needed to see the audience through the eighty-five items scheduled for presentation.[15] Not all entertainments were as substantial as this to be sure, but a long program was considered a decided virtue. The entertainment at the Rock House District, in Douglas County, lasted four hours[16]; one at the Scholten school, Barry County, went on for five hours[17]; and in the Lee District of Lawrence County, "the exhibition commenced at early lamp light and lasted until 12 o'clock."[18] At the graduation ceremonies of the West Plains College,* in Howell County, in 1896 the program went on so long that it was decided that there was no time for the commencement address. Events came to a rapid conclusion as the scheduled speaker, "in a few well chosen words," presented the graduates with their diplomas.[19]

*A number of small "colleges" dotted the Ozarks, offering courses of little more than high-school level. They were private institutions, usually operated by the "professors" who owned them. They provided their students with "cultural" courses, primarily music and elocution.

"Saying her piece" at an Ozark closing-of-school program for classmates, proud parents, and patrons. Courtesy State Historical Society of Missouri.

The schoolhouse was often elaborately decorated for the occasion. Bunches of wildflowers, "garlands of evergreens and mottoes," and "beautiful pictures" were commonly scattered about. The Lone Star correspondent described a Barry County school: "They decorated the school very artistically with evergreens, flags, red, white and blue paper chains, large pictures of famous men, beautiful wall mottoes, and lovely paper flowers of all colors."[20]

At Rogersville, in Webster County, it was noted that the stage was elevated and the entire schoolhouse was tastefully decorated.[21] The stage was probably a temporary one erected for the occasion. A Greene County schoolteacher proudly described such a stage in her personal diary, suggesting the labor and ingenuity necessary to present a satisfactory end-of-school program. She wrote: "Was carpenter boss today. Had my boys build a stage 8 x 16. Trees are so very plentiful they cut a big tree in the school yard and sawed it into blocks with cross cut saw which they covered with oak lumber. It is a dandy." To complete the stage, she took up a collection from her students for material for a curtain, which they fastened to a wire with small harness snaps.[22] Guy Howard concluded, as he watched the wagonfuls of patrons arriving for a program, that "evidently the whole country had taken a holiday in deference to the school closing."[23] Newspaper reports indicated the great popularity of the closing program. If the house was not "filled to overflowing," the correspondent seldom allowed that fact to see print. The boast that at Rock School there were "many standing and a large number driven away on account of their inability to secure even standing room"[24] was typical of attendance claims.

A Taney County patron was deeply moved by a patriotic program: "Some grey heads bowed down when the lovely poem of The Challenge with its touching references to the blue and the grey, Dixie, and Yankee Doodle was delivered by Mr. Owen McClary, and old hearts beat time again to the thrilling incidents of Sheridan's ride."[25] As sentiment touched one viewer, humor appealed to another. The program that he saw was long, "about sixty pieces consisting of the grave, gay, comic, serious, pathetic, and ludicrous." His highest praise was reserved for the "roaring farce, *The Love Sick Darkie,* which closed the literary exercises and also came near doing as much for some of the audience." He was particularly fond of one scene: "When the darkie took poison and was so wonderfully cured by the doctor the scene beggared description and many of the audience was in the queerest shapes they were in since childhood."[26]

Not all the programs were as successful. A patron who attended the closing exercises at Pleasant Hope in 1899 became so disgusted by the many disappointments of the evening that he made no effort to conceal his unhappiness. Although the house was overflowing with patrons and visitors, the exercises did not start until nearly nine o'clock because the music was late. Then it was discovered that the only violin was broken, and the instrumental music had to be canceled. Even the debate and the play came in for criticism:

> Two of the speakers were absent and had it not been so thoroughly announced the debate would have been abandoned for it seems as if the negative were depending on the absent speakers and did not care whether they prepared or not. . . . The play was begun and played under very adverse circumstances. The debate did not end until ten o'clock at which time the audience was weary. The stage and curtains were not sufficient and we were informed that the players had not practiced since they played at Bois D'Arc.[27]

Entertainments in the town schools—whether public,

high school, normal, or college—were sometimes attended by the local editor, who often demonstrated his critical prowess in a detailed analysis of the performance. On June 18, 1896, the editor of the *West Plains Journal* devoted three columns of his front page to a review of the commencement exercises of the West Plains College. He saw a parallel in history between the assassination of Lincoln and Garfield "in our own day" and in the college presentation of "The Coronation Pageant of Annie Bolyne, whose tragic death made the coronation with its pomp and glory but a mockery, and history repeats itself." The girl who recited "Poor House Nan" had a voice that was "clear and distinct" and "reached all in the room." "Some very fine recitations and declamations are often spoiled," he observed, "because often delivered in so low a tone only those near the stage can hear."

The final item on the program, an unnamed farce, led the editor to a digression on the topic of woman suffrage:

> The acting was so real several ladies in the audience drew their feet up from the floor and contemplated getting up on the chairs, and were only deterred from doing so by the thought of how those horrid men would laugh and say, "She wants to vote and is afraid of a mouse," but why this fear of a mouse should disqualify her for voting we fail to see.

The reviewer of the closing exercises of the Ash Grove school in 1895 was not altogether enchanted with the operetta, *The Enchanted Wood:* "The girls were not in the best of voice and an apparent stage fright to some extent detracted from what they were really able to do."[28]

No reservations were expressed about a school entertainment at Galena, where "there was not a shadow of a failure upon the part of any of the participants." The critic believed that the best thing on the program was the Tom Thumb wedding. Small children were dressed in appro-

This Tom Thumb wedding took place in Poplar Bluff, Butler County, on the eastern edge of the Ozarks. Courtesy State Historical Society of Missouri.

priate costume—the bride in white, boys in black swallow-tail coats, and the minister in a long white robe: "The little people went through the trying ordeal in a way that won the heart and admiration of everyone present. The polite and gentlemanly bearing of the little boys and the sweet charming manner of the little maidens was indeed a lesson to those much older."[29] A Barry County editorial of 1898 excoriated such performances. The editor wrote, "When one reads of . . . mock marriages between tiny children at charity affairs, one better comprehense [sic] the growing levity with which solemn ceremonies are invested."

Theodore Carter, in an oration at a West Plains High School commencement, so vividly described the last hour of Benedict Arnold's life that a correspondent to the *West Plains Journal* of April 23, 1896, was inclined to "pity and forgive him" even though "there is no one more despised and execrated than a criminal." An Ash Grove reporter was impressed by Triece Chandler's optimistic approach to the topic, "Morals and Politics": "His oration showed careful preparation and considerable maturity of thought. His delivery was easy and natural and with proper training he will develop into a fine speaker."[30]

"Delsarte drills" were regular features of most closing exercises, as well as of other school programs. In these pantomimic drills the carefully prescribed movements and positions of the Delsarte system* were executed with grace and precision. The system included eight foot positions, beginning:

> Position 1—Right foot diagonally front and be-yond left foot.

*After François Delsarte (1811–71), a French educator and singer, whose system of dramatic expression was aimed at making elocution a science through training, by fixed rules, of the voice, body attitudes, & gestures.

Position 2—Right foot diagonally back and beyond left foot.
Position 3—Right foot diagonally forward—right.[31]

Also prescribed were arm and hand positions and head
and eye positions, as well as rules for transitions from one
position to another and for coordination of the various
elements.

Delsartean pantomimes coordinated a reciter, pantomimic action in accord with rigid rules of movement, and
musical accompaniment. The reciter declaimed a line of
poetry while the pantomimist performed the prescribed
action:

> Line 7 [From "Flag of the Rainbow"]—"While but
> a man is alive to defend it."
>
> Position 5. Transition—right hand in first to left
> side, then, by straight line movement, to upper zone,
> oblique front, right, thumb part of hand upward as if
> holding a sword, left hand clinched in lower zone, out
> at side, face strong, whole manner earnest and enthu-
> siastic.[32]

Both drills and pantomimes were popular closing-of-
school events, as indicated in this item from an Ozark
County paper:

> Standing room was at a premium at the school
> house. . . . The drill of Mrs. Platt's class in Delsarte
> by 10 of Gainesville's most comely young ladies is es-
> pecially worthy of mention, every movement showing
> grace attained only by careful and painstaking study.
> More credit is due when it is known that they had but
> about 10 days practice.[33]

Other drills, perhaps inspired by though not necessarily based upon the Delsartean system, found popular
acclaim with audiences. The patriotic fervor of the Span-

ish-American War was still evident in the drills given by the pupils at Ash Grove in 1899: "They were dressed in the stars and stripes and as they marched the stirring strains of 'The Red, White, and Blue' burst from their young lips on the ears of the audience."[34] The Normal School commencement at Gainesville in 1898 featured

> first the broom drill by sixteen young ladies in beautiful costumes of navy blue with white braid. The evolutions of their march and manuel of arms would have done credit to old veterans. Also the young men's drill and song, "Marching Through Cuba" was well executed and appropriate. "Last but not least" was the little girls fan drill and good night.[35]

Three months previously on the same stage during the good-night drill following a literary program, one of the tiny maids' gowns had caught fire from a candle. It was, fortunately, "soon extinguished by Prof. Platt and Grandma Conklin who happened to be on hand at the right time."[36]

End-of-school audiences applauded the production of plays bearing such titles as *The Deestrict Skule, The Old Maid's Triumph, The Bank Cashier,* and *Our Country* ("full of life and patriotism and beautifully illuminated with tableau"), as well as the old standbys *Uncle Tom's Cabin* and *Ten Nights in a Barroom.*

"Sweet songs by the Plum Valley singers, sweet instrumental music of high order on three instruments, and fine order, added pleasure to the occasion" at the Plum Valley school,[37] and a Texas County district pledged money to "secure the services" of a Graphophone, "which travelled twenty four miles to visit."[38]

After the entertainment the final treat of the day was produced from its hiding place by the schoolmaster. The distribution of candy to the pupils and to all the children in the audience was an invariable ritual, and one eagerly awaited in a region where store-bought sweets were rare indeed. Following dismissal, most patrons seemed to feel as did the correspondent who wrote: "To tell it in as few words as possible, the whole affair was a decided success and enjoyed by all present. There were people present who laughed out loud who haven't smiled before for years."[39]

While we had a place of a religious nature to go we cared not for the dance nor had we any. Now we have no opportunity to attend religious gatherings, consequently Satan's road is left open without impediment. Let us again have the blessings of religious meetings.

— HOUSTON (TEXAS COUNTY) HERALD, SEPTEMBER 12, 1901

Religious Gatherings

THE good people who inhabited the Ozarks hills were predominantly Bible-believing, traditional Protestants, who took their religion seriously but seldom somberly. In spite of their freedom from the concentrated vices of the big cities, Ozarkers were strongly aware of the presence in their midst of the ever-zealous devil and were equally zealous in their efforts to rid themselves of him and his manifestations of evil. Evidences of the "old gentleman's" efforts to corrupt were seen principally in certain amusement activities and, of course, in the intemperate consumption of alcoholic beverages.

Whiskey was the devil's brew, and by whatever name it was called—"moonshine," "blue ruin," "old bust head," or "mountain dew"—it was blamed for being, and probably was, at least partly responsible for most of the disturbances, fights, killings, desertions, and depravity in the Ozarks. But the hillfolk saw the devil's handiwork in other activities as well. An Ash Grove College writer described the dangers of a "show": ". . . Satan's favorite way of exhibiting his power over the sons and daughters of men, chucking them under the chin, snapping his fingers in the face

The Vera Cruz Baptist Church. Vera Cruz (or Veracruz), the first county seat of Douglas County. The county seat was moved to Ava in 1870. Courtesy State Historical Society of Missouri.

of the church, and dismissing schools against the wishes of teachers and the decrees of school boards.[1]

Card playing was, in some communities, a legal as well as a moral crime. "Several lawyers" admitted to playing cards on Sunday in Douglas County and were fined a dollar apiece for the offense.[2] The minister's alliance of Springfield passed resolutions against playing with the "spoiled pasteboards," which activity was condemned as being, at best, "but a process of killing time which, like the destruction of life, is a grievous sin against God." The ministers saw this activity as a direct path to lawlessness: "There is many a man behind prison bars as the result of being impelled to criminal associations and to crime itself by the gambling spirit nourished in his soul around his mother's card table."[3]

In many sections of the Ozarks dancing was viewed with a disfavor ranging from disapproval to vehement condemnation. About 1910 a man and his wife went to a small church in Greene County and delivered anatomical lectures illustrated with stereoptican slides. Addressing the shocked and embarrassed women and their daughters who had heard the lecture on male anatomy, the woman lecturer concluded, "So girls, you see what's pressed up against you when you dance with a man."[4] A Barry County preacher rejoiced that, in spite of "much opposition from Satan such as dances in the same locality," his three weeks' revival was successful and resulted in eighteen converts.[5] "Papa's Joy," the correspondent from Stultz, in Texas County, felt that "play parties, shindigs, and hops" were not very "elevating" to a community and were, as a matter of fact, "very degrading."[6] Another Texas County writer was "surprised to think that parents that claim to be christians [sic] will allow their girls to go to such places [hoedowns]. Boys, what can you promise yourselves in taking these girls for wives?"[7] And they were of Texas County stock, those young ladies who walked several miles home

70

after they were taken by their escorts to a dance instead of to a reunion as promised.[8]

Perhaps it was because of its association with the dance that the fiddle acquired its nickname "the Devil's Music Box." Ozarkers who would never consider participating in a square dance where a fiddle accompanied the couples through the figures of a dance such as "Skip to My Lou," would go through the same movements at a play party while the participants sang the words without accompaniment save the clapping of hands. The addition of a fiddle, or, for that matter, almost any other kind of instrumental music would have converted this innocent "game" into a wicked "dance." Guy Howard tells of the old lady who arose in church after a young girl had played a sacred number on the violin and in righteous indignation exclaimed: "Lord have mercy that I should live to see the day that arch instrument of the Devil, that dancin' fiddle of sin and shame, should be brought into the Lord's house."[9] It is interesting to note that, while most churches condemned dancing, kissing games were often popular adjuncts to proper play parties.

Ozarkers looked to their church to help them identify and combat the Devil's activities on the earth for the good of their own souls and for the social and moral betterment of their community. Their religion was sincere and fundamental, and it provided them not only with nourishment of the spirit but with recreation for the body as well. Religious services provided an opportunity for a community to gather in social fellowship, and the opportunity was seldom missed: "The people of New Site, always bent on having a good time, have been attending all the camp meetings, picnics, and so forth for miles around."[10] The lack of religious meetings in a community was seriously felt as a social loss, especially by the younger generation: "Jack's Fork is dead. Quite dead. The young people are about to lonesome to death. Such a pity the church can't be finished so we can

71

have prayer meeting, Sunday school, and preaching."[11]

Claude Hibbard commented that "the church was your outstanding social function because there you met all your neighbors that you hadn't seen for a week." Chick Allen, waving his arm across the Stone County hills and indicating a distant ridge, explained why the Ozarkers considered their churchgoing so important as a social occasion: "Well, maybe you lived over there five mile and I'd see you once't a week. We'd meet on the road and I'd throw my hand up at you but we couldn't set and visit. Well back in that day and time, visitin' and good fellowship with your neighbor and your friend, that was *it*! We didn't have places to go."

Of the religious gatherings in the Ozarks regular preaching and protracted meetings were the most popular, but the Ozarkers also enjoyed attending children's-day exercises, religious debates, and baptisms.

PREACHING

Many communities in the Ozarks lacked a regular minister, regular church services, and even a church building. If there was no church, the local schoolhouse was normally utilized for that purpose. State law required that the taxpayers of a district vote approval for use of the schoolhouse for purposes of worship.[12] Such approval was usually forthcoming for the taxpayers and school patrons were also the churchgoers. An interesting exception occurred in the Elsworth community, in Texas County. No religious services had been held in the community for several years, and in a letter to the *Houston Herald* the writer indicated why the use of the school had been denied for church purposes:

> Was it not while church was being held in the school house that Mr. Lawrence Ross got his saddle cut up? Not to say anything about Mr. Petterhoff's saddle and he preaching for the people. Was it not

72

Bert Sallen's horse's tail that was sheared? . . . Was the
flue of the school house not torn down while church
was being carried on in it? After the people voted
church out of the school house, did not the church
come and break open the doors of the school house
and go in there and last but not least was the school
house not burned and who did it?[13]

Such an instance was the exception. Most communities
were pleased to offer the schoolhouse, or whatever other
facility was available, for religious purposes.

Few rural communities were large enough or prosperous enough to support a full-time minister. The preacher
held services every second or third or fourth or fifth Sunday, visiting other communities during the intervening
weeks. Most country ministers did not depend for financial support upon their religious vocation for their liveli-
hood but pursued other occupations, such as farming and
schoolteaching during the week. Many wanted no money
from their ministry and would accept none, reasoning
that since Christ and his disciples had received no pay,
preachers should ask for none.[14] Others assumed a more
worldly posture and believed that they were entitled to be
supported by their congregations. A minister on the Bois
D'Arc–Billings circuit termed himself "the cheap preacher,"
a title arising, a correspondent explained, "according to
the way in which he was reimbursed quite recently."[15]
The pastor of the First Baptist Church at Ozark was less
philosophical about his pittance. He gave his people a
scolding for their lack of support and left them forthwith.
A correspondent writing to the newspaper about this incident further excoriated those who came Sunday after Sunday to be "entertained" in the church but did not give of
their means to it.[16]

Many communities were disappointed by preachers
who failed to keep their appointments. When one finally
appeared at the Pleasant Grove Church, in Texas County,

Saint Peter's Evangelical Church, in Billings, Christian County, decorated with Christmas trees, 1910. Billings had a large number of German settlers. Courtesy State Historical Society of Missouri.

for the first time in more than two months, he found only a small congregation waiting. The correspondent explained: ". . . it has been so long that people did not turn out as we generally do, fearing they would get left again. What's the matter with the preachers on this circuit anyway."[17] A writer from Aix (Barry County) offered "a scarcity of grub" as a possible explanation for the wide berth that preachers gave his community. He pleaded: "Won't someone take pity on our benighted condition? We can promise them peas and potatoes and a few such luxuries as greens and gooseberry pie."[18]

A correspondent rejoiced at the offer of a Eureka Springs minister to hold a meeting in Cassville. "For years we have been left during the winter time to wander in heathen darkness," he commented sarcastically, "and only when spring chickens, sweet potatoes and other luxuries were the bill of fare could we hope for a ray of light."[19] Although the situation was general throughout the Ozarks, Barry County seemed to be particularly afflicted with unreliable ministers. "There have been two thirds of the appointments missed here in the past year," complained a Butterfield correspondent to the *Cassville Republican* on July 8, 1897. He was hoping, he said, for some missionary-inclined preacher "who is going about doing good" to fill his appointments or explain the reason why he did not. Two weeks later the same writer reported that another preacher had failed to appear and sighed resignedly, "This is getting to be the worst world for disappointments we ever lived in anyway."[20]

In all fairness to the ministers of Barry County, it should be reported that there were those who did meet their appointments only to find that the congregation had failed to meet theirs. An Ash Grove correspondent reported: "Elder Wood filled his appointment Sunday, . . . but all his brethren left on Friday or Saturday before for Cassville to the Free Silver Rally and left him with empty benches to preach too [sic]. The white mettle [sic] here is at white heat. It beats religion."[21]

When the preacher arrived in a community to hold services, people of all faiths gathered to hear him. Most hill ministers preached a fundamentalist message that was free of sectarian spirit and that "all Christians irrespective of creed or denomination could endorse."[22] Polemic issues were left to be settled in extended religious debates and were usually pointedly not discussed in such interdenominational meetings. Instead, the preacher expounded upon certain biblical texts; exhorted against wickedness, worldliness, and intemperance; and painted vivid word pictures of the miseries of sin upon the soul and the suffering of lost souls in hell. He was often equally adept at depicting the rejoicing of the redeemed at the return of the sinner to the folds of righteousness.

Sermons were long, sometimes continuing for two hours or more. While many of the preachers were educated to some degree or another, many could not read the Bible from which they were preaching. Chick Allen explained: "There was a world of them preachers that couldn't even write their name, but they would maybe have somebody to read their text you see, and otherwise they would just get up there and take off. Otherwise it was just them and the Good Lord, that's all." Otto Ernest Rayburn wrote that "the hill preacher does not talk from notes or prepared manuscript, but depends almost entirely upon the inspiration of the moment."[23] Many preachers prepared sermons on a particular text, which they used over and over again, even before the same congregation. Mrs. Dunlap recalled that young people of her community used to say, as they prepared for the monthly trip to church, "Well, let's go hear about Lazarus and the rich man again."

A minister who gained a reputation for eloquence in the pulpit soon acquired a loyal group of admirers, and his sermons were enthusiastically reviewed by parishioner-correspondents:

> Rev. J. A. Brown delivered a masterful discourse at the Coates M. E. church last Sunday evening to a large and appreciative audience, using for a text Luke 15, verses 18 and 19. Mr. Brown was at his best and painted a word picture of the miseries of sin upon the soul and closed his discourse by showing the return of the sinner and the great rejoicing of the redeemed over his return. Those present were deeply impressed and pronounced it a spiritual success for good.[24]

> Parson Moorhead gave us Boredarkies [residents of Bois D'Arc] a rousing sermon Sunday. He knows just how to hit the old sinners, I mean the old sanctified that think they are going to Heaven on flowery beds of ease. He is now at Ozark telling them of their faults and trying to get them to reform.[25]

Other correspondents described the performances of their favorites in equally glowing terms: "The most brilliant and oratorical sermon that it has been our privilege to hear"; "An interesting and effective talker, startling at times, but always commanding attention"; "One of the grandest sermons ever heard in this community"; "Unique specimens of homilitic [sic] arrangement and delivered without ostentatious display." Through these self-proclaimed press agents many Ozark preachers attained local star status.

REVIVALS

Each year every Ozark community was treated to at least one theatrical event in the form of a revival, or "protracted meeting." Visiting evangelists who were invited to conduct the services were commissioned to lead the way in "tearing down the mighty bulwarks of his Satanic majesty."[26]

A successful meeting meant that "some of the wickedest sinners were converted and repenting sobs were brought from those who were never known to shed a tear,"[27] and the community basked for months in an afterglow of remembered fellowship, emotional stimulation, and spiritual satisfaction.

The crowds who jammed the meeting houses had come eager not only to hear the message of salvation and to be reminded of the dangers of eternal hellfire but also to partake of the excitement of the spectacle, to witness the theatrical performances of the preachers, to observe the emotional displays of others, and to abandon their own emotions in full empathic response to the dramatic climate of the occasion. These elements, plus the joy of renewing old acquaintances and the stimulation of a break in the monotony of everyday living, made the protracted meeting one of the Ozarker's major entertainment features of the year.

Emmett Yoeman described revivals candidly:

> Revivals? I think "recreation" would be a pretty good term to describe them. It was not nearly so much spiritual as it was recreational. It was given over to a lot of loud preaching without much thought, long talks without any planning, and lots of singing and lots of shouting, lots of testimonial, and it [the sermon] was a long-drawn-out affair, lasting maybe two or three hours. I would say that it was as much a matter of entertainment and was devised for that purpose more so than any spiritual purpose.

A notice in the *West Plains Journal* of July 22, 1897, announcing a revival meeting at a local church, promised, "You will be entertained, you will be instructed, but can [*sic*] but be benefited."

"I think we would agree that a religion without emotion is no religion," mused Claude Hibbard, "but that a religion that is all emotion is a very poor religion." Certainly the Ozarkers' religion was imbued with a liberal amount of emotion, which also made these religious events attractive as entertainment that they were free to enjoy in full clarity of conscience.

Regular revival meetings were held by nearly every congregation in whatever building was serving them for a sanctuary, be it a schoolhouse, a meetinghouse, or a lodge hall. The seating capacity was limited by the size of the structure, and the meetings were usually associated with the denominational label of the church holding them, though frequently the churches of a town would cooperate in sponsoring union revival services. The most colorful and best-enjoyed protracted meetings were those held in a special place—a brush arbor or a tent—to which citizens of all faiths flocked from miles around in covered wagons loaded with supplies, prepared to stay until the last soul was saved. These were the camp meetings. Uncle Joe Cranfield said over and over, "I'd give a hunnert dollars right now to go to a real old-fashioned camp meeting for thirty days!" The typical meeting that Uncle Joe longed for was described by the State Historical Society of Missouri:

> Preaching usually began at ten in the morning, and if there were three preachers, and a good camp meeting always had at least three, they divided their time so that one spoke in the forenoon, another in the afternoon, and the third at night. A pioneer preacher was a powerful man, strong of lung, strong of body, and a master promoter of his faith. While the people sang or shouted, he walked up and down, urging them to "jine" the church. By the time the camp meeting was in its second or third day, it excelled any circus in interest. Sometimes the minister held his audience spellbound for two or three hours while scores dropped exhausted to the ground. Exhorters by the dozen would shout in unison. A young girl in the front row

79

would get the jerks, swishing her head back and forth in vigorous, rhythmic motion. An old pioneer from the river country would try to shin it up a tree, to get as near to heaven as possible. Sometimes the preacher would have to stop his sermon, get down from his pulpit, and help chase some drunken rowdies away, for bootleggers always hung around the camp-meetings with their jugs and cups, and many a scoffer was made over-bold by another kind of spirits. . . . Next morning after breakfast, the preaching was taken up again with renewed vigor.[28]

Camp meetings were usually held during the latter part of July or in August, after most of the crops had been laid by, and the farm family could take a breathing spell. No respectable camp meeting lasted for less than a week, and if the weather was favorable for living outdoors and the "spirit was in the meeting," it might last for two weeks or even a month.

Campers who came from some distance away brought plenty of provisions and counted on remaining for the duration of the meeting. As Uncle Joe Cranfield described it: "Some old farmer off maybe twenty miles would load up his wagon with grub and horse feed and drive in there, and he'd stay till his grub played out and he'd go back home and get another load. . . . Maybe he'd go back and get a second and a third load." Some families pitched tents or improvised shelters from wagon covers or quilts and blankets for protection from sun and rain.[29] Others lived in their covered wagons. At Chesapeake, in Lawrence County, a "boarding tent" was provided on the grounds for the accommodation of visitors from a distance.[30] Sam Miller mentioned that thirty or forty small cabins were built at a campground in Wright County.

Some communities established permanent campgrounds, which became the site for annual religious gatherings. One of these, at Ebenezer, in Greene County, belonged to the Southern Methodist church. A building on the grounds was described by L. O. Wallis as a "permanent shed, wasn't boxed up or anything, but had a shingle roof and was built good and strong." Before the time the meeting was to begin, the faithful hauled in several loads of straw and piled it all over the dirt floor of the shed. Rough two-by-twelve-foot boards served as pews, and an eight-by-ten-foot platform elevated the pulpit, which, together with the piano or organ, was brought over from the church building. There were no boarding facilities at Ebenezer. Those from a distance camped or lived in their covered wagons or stayed with friends nearby.

At Silver Shade, in Douglas County, "a large, commodious shed thirty by fifty feet" housed the meeting and there were an abundant supply of cold spring water and plenty of free pasture for the horses, on land enclosed by a wire fence.[31] Other meetings advertised similar facilities for the convenience of the campers.

Many communities preferred to follow the traditions of early settlers and conduct the services in the open air, sheltered only by an arch of trees and foliage known as a "brush arbor." Uncle Joe Cranfield described the construction of one of these shelters: "The men would go out and hunt a place out in the woods, . . . and they'd cut poles and forks and build a scaffold up overhead there and then they'd cover that with brush for shade. I've seen as much as an acre covered that way." The brush arbor would not protect the congregation from a pouring rain, but it did provide shade from the blazing August sun, and at night when the preacher exhorted the worshipers to "get right with God," somehow, under the stars and in the open air, He seemed just a little bit closer.

One minister was usually in charge of the camp meeting and remained in residence for the entire period. He was invariably assisted by several other men of God during that time. "I've seen as many as forty or fifty preachers at one of them," said Uncle Joe Cranfield. "Every preacher around in the whole country would come. It didn't make any difference about his denominational standing. He was 'Brother So and So.' . . . They left off their denominations. They preached the Bible."

L. O. Wallis remembered a minister who happened to be passing through during a camp meeting and remained to speak at several sessions. A meeting at Arno, in Douglas County, promised that "several preachers from a distance will be in attendance."[32] In Barry County, "A brigade of preachers about the same time attacked the camp of the old gentleman's [the devil's] soldiers . . . and routed them, scouted them, nor lost a single man but took some prisoners."[33]

The earlier description of the typical camp meeting suggests something of the style of preaching typical of the services. A minister never received praise for subtlety and tact. "A man who is so plain and forceable [sic] in his way of talking to the sinner that they can't resist his persuasive voice"[34] was an accolade for a preferred approach. L. O. Wallis recalled: "They really preached hell fire and brimstone. . . . There was no compromise in their preaching at that time. They didn't pull any punches." Fred Steele observed, "There was more arm waving. People liked this. It was what they expected."

Many revival ministers preached in a style that has been described as a "revival monologue,"[35] dramatizing their sermons by adding imaginative conversation and pantomime to make vivid a text and to add interest and emotional appeal. One revival preacher

> carried us back to the Flood, running across the rostrum to pound on the wall (the door of the Ark) as he screamed, "Noe-y! Noe-y, let me in! Let me in!" But Noah refused to hear his plea. He dropped to his knees, fighting the raging waters around him, and

then, arms wide, still on his knees, he made the plea
that all who were not uneasy for their souls come
knock at the Ark to which Jesus would gladly admit
us.[36]

Revival preachers encouraged audience participation,
arousing with their dramatic sermons and emotional ap-
peals shouts of "Amen!" and "Go it, brother!" L. O. Wallis
recalled: "I've seen fifteen or twenty people shouting at
once. Sometimes during the preaching or . . . during the
service following the preaching, when they'd call for penitence and so on, or maybe sometimes it would pretty near
break up the service there for a little while during the
preaching." Preachers found it rewarding to sprinkle their
sermons liberally with humorous anecdotes or homespun
jokes, for an audience that laughed was a responsive audience.[37]

One of the outstanding features of a camp meeting
session was the daily testimony service. People stood to
tell of their religious experiences, and what the Lord had
meant to them in their everyday lives. As Uncle Joe Cranfield put it, "You tell what you been a-doin' and how you
felt about it and what you wanted and what the Lord had
done for you." He described a testimonial meeting that
started one night a half hour before preaching was to
begin:

> There was an old lady got up from Arkansas. She
> started in. She said, "When I first went to Arkansas,
> there was no Christians down there that I knowed of
> but myself." Says, "I finally got acquainted with a
> neighbor woman, a Christian. She belonged to my
> church." Says, "We'd get out and meet in the woods
> and have prayers. That was about all we could do,
> and kiss each other when we'd separate." . . . She
> says, "And it was like pulling water over a pulley.
> We could feel that spirit come down."
> That thing broke loose, and they got that testi-

mony meetin' started, and it lasted till after two o'clock, and they was thirteen conversions came out of that testimony. The preacher couldn't do nothing with 'em. He tried to stop them, but they wasn't nothin' doin'. They was testifyin' all over that place.

A revival meeting was counted a success when it resulted in a large number of conversions. Newspaper accounts proudly reported: "44 additions to the church up to the tenth day of the meeting,"[38] "shouts in the camps, twenty souls . . . saved to God,"[39] "more than 200 conversions at a big protracted meeting at Ava which has been going on for several weeks,"[40] "an 18 day meeting . . . with 43 additions."[41] The plain-spoken correspondent to a Christian County paper rejoiced in the eighty "accessions" to the church during a protracted meeting:

> There was quite a number of accessions from other churches besides those who had heretofore joined and then danced themselves out of the church but came back to the fold. A few old toughs like Don Wilson, W. M. Randolph, H. W. Stewart, Alex West, Ab Stiffler and Mrs. Amos were all that made their escape and they had to hide out.[42]

The seventy-two conversions at Jerico (Wright County) in 1887 included the town's two saloonkeepers. These gentlemen turned over their liquor stock to the ladies of the WCTU, who proceeded to burn the liquor in a grand Saturday-night bonfire in the middle of town.[43] That same year Wright County had voted itself wet in a local option election. In 1905, however, the voters overwhelmingly approved shutting down all "dramshops."

A sad note in the midst of such general rejoicing is found in the *Ozark County News* of November 19, 1896: "Henry Jenkins of the southwest part of this county is deranged. It is supposed excitement over religion is the cause of his mind losing its balance. He attended two protracted meetings in succession, making several weeks since which time he has been entirely crazy." Considerably more cheerful, and certainly more typical of the general reaction to revivals, was the editorial comment in the Ash Grove paper at the end of a two-week union evangelistic meeting. The writer believed that the cooperative effort aided in "breaking up a sectarian divisive spirit in our people which has necessarily hindered every public enterprise in this community." The evangelist, in his farewell remarks, added his plaudits: "This is the only meeting of two weeks that I have ever held where I have not heard an oath or seen a drunk man."[44]

CHILDREN'S DAY EXERCISES

During May, June, and July many Ozark churches presented their Sunday-school pupils in a recital-program called "Children's Day." Like many other Ozark entertainments, Children's Day provided an opportunity for friends, neighbors, and relatives to gather for fellowship and entertainment.

More than five hundred people gathered in a grove at Merrit's Spring (Douglas County) in 1889,[45] and an estimated crowd of one thousand was on the grounds at the Phenix Children's Day in 1892.[46] In 1894 visitors arriving by train to attend the exercises at Miller (Lawrence County) were furnished transportation to and from the church.[47]

The programs of most Children's Day exercises consisted of religious, sentimental, patriotic, and temperance performances by the children. The following program, printed in the *West Plains Journal* of June 9, 1898, is representative:

Forenoon

Processional March
Opening Chorus

Responsive Scripture Reading
Prayer by the Pastor
Chorus—"Hail the Day"
Recitation—"Angel of the Flowers"
Recitation—"Mother Earth"
Recitation—"Sunbeams," "Rain and Dew"
Song by Helpers—"All is Complete"
Recitation—"Dandelion"
Chorus—"Brave, Honest, and True"
Recitation—"Appleblossoms," "Johnny Jump-up"
Song—"Don't be Cross and Surly"

Afternoon
Recitation by Infant class
Duet—"Heartease"
Recitation—"Waterlily"
Chorus—"Be Pure in Heart"
Daisy Chain Marching Exercise
Recitation—"The Golden Heart of the Rose"
Closing Chorus—"The Golden Heart of the Year"

In 1889 the children's recitals at Little Beaver, in Douglas County, were supplemented with addresses by a county judge and an elderly settler,[48] and Miss Mattie Moore spoke "one of the most touching pieces of the evening," when she recited "In a Tenement House" at the Marshfield (Webster County) Children's Day in the same year: "She did herself great credit, also her manner and delivery and general appearance was evidence of one or two things; great natural abilities or hard study upon the part of herself or good training by some one."[49]

Many of the orations and recitations stressed the temperance theme. "They were enough to turn the hearts and minds of all who heard them against the use of intoxicating liquors,"[50] wrote one reporter, and another prayed, "May there not be a one of all who listened who will sink down so far in darkness and misery as some that was pictured out so plainly to us."[51]

86

The churchhouse or arbor in which the program was held was usually decorated for the occasion, and the children wore their very best clothes. Mrs. Earnest Hair recalled that she preferred Children's Day to Christmas because the former always meant a new white dress. At a program in 1898 in Howell County the choir was colorfully arrayed in "badges of red, white, and blue, and the spirit of patriotism was manifested even in the very small children."[52] The Cumberland Presbyterian (C. P.) church in Mount Vernon was festooned with "birds in cages and an abundance of flowers,"[53] and the black children of Marshfield built a cross of different-colored flowers to the accompaniment of music and singing by the children.[54] A description of the Moody Church Children's Day stage appeared in the *West Plains Journal* of June 30, 1898: "The arch at the front of the stage was a very attractive feature in the decorations. 'Children's Day' in large, cedar-covered letters intermingled with roses covered this arch and at the foot of the stage a profusion of lovely plants and flowers. Flags were seen waving from almost every direction."

With dinner on the grounds; singing, speeches, and recitations by scrubbed and costumed children; and a colorfully decorated setting for the performance, Children's Day was a favorite springtime entertainment in the Ozarks.

RELIGIOUS DEBATES

The Ozarkers' love of a good argument found considerable gratification in the religious debates that were periodically staged, usually between two ministers of differing religious beliefs. Points of doctrine and theology that were not ordinarily addressed in the fundamentalist sermons preached by most hill ministers received full consideration in these heated polemic discussions.

A very popular subject was proposed for debate at

Gainesville (Ozark County) in 1904: "Resolved: That water baptism is essential unto salvation."[55] In a ten-day debate in Greene County between a "Campbellite" and a Methodist minister the topic of baptism was divided into four parts: (1) "Is immersion in water the baptism prescribed by Christ and the Apostles, or will some other do just as well?" (2) "infant baptism," (3) "baptism for the remission of sins," and (4) "The Holy Spirit in the conversion of a sinner operates directly or immediately on the heart."[56]

"A large and orderly crowd was in attendance" at a debate on baptism held at Northfield (Lawrence County) in 1886.[57] While a similar debate at Washburn (Barry County) in 1895 was well attended, a correspondent feared that in the last analysis "the disputants were more firmly set in their respective opinions while the minds of the audience were unchanged."[58] A minister of the Christian church and one of the C. P. church engaged in a debate on the same topic, "something which in all probability neither one of them knew much about . . . and which as a rule generally results in each side swearing by its man and stirs up sectarian bigotry in the community and gets neighbors to hating one another."[59]

Religious debaters often sought to establish the authenticity and antiquity of their particular denominations through such propositions as this one, debated at Ash Grove in 1891: "'The Baptist Churches with which I, Elder James Bandy am identified are the churches of Jesus Christ.' James M. Bandy affirms, Dr. Lucas [a Disciple of Christ] denies."[60] Interest in this debate ran very high. It began on a Monday morning and continued with two sessions a day until the following Saturday afternoon. People came for miles around, and good order was maintained throughout. A crowd overflowed the schoolhouse at Taneyville in 1896 to hear a debate on the historical origin of the Disciples of Christ (Christian) church, the negative asserting that it began with Thomas and Alexander Campbell.[61]

Representatives of the Church of Jesus Christ of Latter-Day Saints were a particularly argumentative lot, taking on any and all comers on the question of the identification of their church with the one established by Christ, though they seldom received much public support. Five hundred people heard a discussion by Christian preachers and Mormon ministers in Douglas County in 1893. In a vote taken to determine which side had the best of the argument, the Mormons were overwhelmed 487 to 13.[62] E. H. Vandiver recalled a week-long debate between two ministers, a Methodist and a Mormon, named Black and Blue. He marveled that "both boarded where I did and they never fell out that whole week," though, in his words, "Each tried to show the other where his wagon was broke down."

The ability to quote Scripture liberally and with facility and to draw forth the emotions of the crowd was apparently requisite for a successful religious debater. A disgusted letter writer complained to the *Houston Herald* that a minister debating the negative on the topic "Resolved: That Christ is a Myth" did not answer the "evidence of eminent historians" presented by his opponent but resorted to "text preaching, appealing to the sympathies of the people, telling funny stories and exhorting his hearers and his grey haired opponent to turn from their sinful ways, etc. etc."[63] Another correspondent hoped that the outcome of a four-day debate might be "that many have learned scripture," for both men were "able debaters and well versed in the scriptures."[64]

Whatever benefits these long-drawn-out theological arguments may have had, the Ozarker liked them and encouraged them for they were entertaining. As Mr. Vandiver said, "I'd walk all night to see one of them debates."

BAPTIZINGS

A baptizing in the Ozarks was at once a solemn religious undertaking and an occasion for the gathering of large crowds to witness the spectacle. Otto Ernest Rayburn described an Ozark baptizing as "a colorful affair of mingled emotions, ranging from deepest reverence to ribald curiosity."[65] Grover Denny said: "Why it would just be like going to a carnival. There would be people from all over the neighborhood there and they enjoyed seein' ya. No one was there to make fun at all, they was there for what good there was there."

The event took on a particular significance after a protracted meeting, when the number to be baptized was large. In 1886 twelve to fifteen hundred people watched thirty-six converts undergo baptism in the Spring River, near Verona, in Lawrence County.[66] Uncle Joe Cranfield, a preacher and faith healer, described such a scene:

> Well, they'd take these converts to the river and baptize them and then everything broke loose when that happened. They'd shout, them old women, and the old men, they'd go wild too, a-hollerin' and the tears just a-pourin', not of grief, of joy. Like the Lord said, "Joy unspeakable and full of Glory!" . . .
>
> I've seen 'em break the ice to baptize. Well, now, a lot of people would think they'd freeze to death, and they don't. I never knowed a one to take any cold over it.

Mrs. Dunlap used the term "Warm-Weather Baptist" scornfully, denoting those who were converted in the winter but preferred to wait until summer for the rite of immersion.

The usual order of baptismal services called for the minister to lead the queue of converts down to the river and into waist-deep water, where they stood while hymns were sung and the minister offered a prayer. One by one

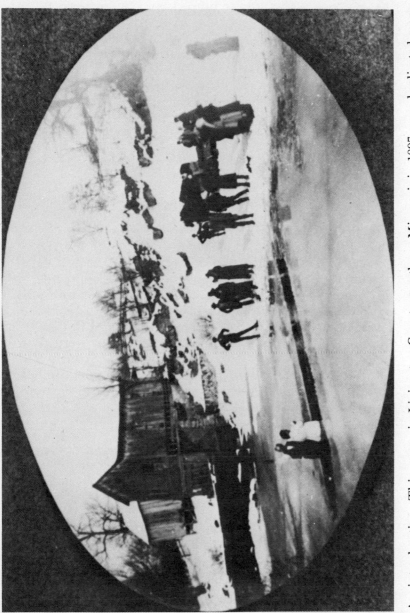

A winter baptism. This scene, in Livingston County, northern Missouri, in 1897, was duplicated many times in frozen Ozark streams. Courtesy State Historical Society of Missouri.

the converts were baptized and then splashed ashore to be greeted by a friend or relative with a large towel or robe (women and girls were properly dressed for the occasion, wearing several extra petticoats and even extra skirts and blouses, not for the warmth they provided against the chilly waters but to prevent their wet clothing from clinging and revealing the slightest hint of feminine form). Another hymn, a prayer, and the service was over.

River baptisms were accompanied by occasional mishaps, most of which had their humorous side. One minister was faced with the task of evacuating from the baptismal pond a reluctant Durham cow who refused to give way to the baptismal party.[67] Another group became entangled in trot lines stretched by fishermen across the stream,[68] and a crowd that gathered at Johnson's Mills (Lawrence County) was astounded when the unrepentant husband of one of the candidates for baptism rode his horse into the middle of the stream and tried to stop the proceedings. "Threats of 'muscular Christianity' finally induced him to leave the spot, and the rites were proceeded without any further interruption."[69] A hog, disturbed when a baptismal party invaded his pond, seized a three-year-old child and dragged it some distance away before it could be rescued. The child was unharmed, save for scratches and bruises. The hog was shot.[70]

A baptism that took place in 1889, though witnessed by few people, surely deserves space in these pages by virtue of its comic-opera.

On May 10, 1889, three men, Bald Knobbers,* were to be hanged in the jailyard at Ozark. As the story goes:

> When the men saw that there was no escape for them from their fate they got religion and decided to

---

*Bald Knobbers were former vigilantes turned outlaws who roamed and terrorized the Ozarks, so called because they often met on mountaintops with no trees, where they could see anyone approaching.

be baptized. The baptismal rite posed a problem for both the law and the ecclesiastical forces concerned, in that the faith to which the convicted had professed required a complete immersion as the mode of baptism. Rites of this nature ordinarily were carried out in a large pond or in a creek, creek water being thought more potent. The jail was innocent of bathing facilities, and after the men's year or more confinement, a clean running stream would seem to have been indicated for the rite, but in this case such procedure was not considered feasible, although it would now seem to have been a practical scheme to have marched the trio down to the creek and while armed guards held their guns on the participants, the sins of the penitents, together with other less psychic and more soluble matter would have been washed away in an orthodox manner. But it was thought best to perform the ceremony in the jail. This required a receptacle that would hold enough water to insure that every part of the anatomy would be submerged at the same time, and it was in search of such a contrivance that the officials discovered that there was a bathtub in town. This was a large and cumbersome apparatus of wood and sheet metal that required it to be filled and emptied by hand. . . . This tub was borrowed and used for the baptismal ceremonies. It had capacity sufficient to permit the penitents being completely immersed, thus assuring him that after his exit from this world where his sins could not be forgiven, he would have a new deal in the next.[71]

The play given by the Thespian club which had been enlisted for the purpose of entertaining the guests during the early part of the evening was executed in a manner that drew comment and praise from all.

—DOUGLAS COUNTY HERALD, MARCH 17, 1887

Local Dramatic Productions

A restless Forsyth (Taney County) citizen seated himself one December day in 1897 and wrote a letter to the local paper, suggesting that, since the nights were becoming long and tiresome, some activity should be inaugurated to pass the time:

> We would suggest that a home dramatics company be organized and give a series of entertainments. . . . It would furnish the residents of Forsyth an enjoyable season of entertainment and also be instructive to the participants. We are so situated in this part of the country that entertainments of this kind is impossible unless we do something of this kind.[1]

Throughout the Ozarks dramatic clubs were formed to produce plays for the purposes noted above, benefits to the participants and entertainment for the audiences—and entertained the audiences certainly were. An "amusement loving people" who needed "entertainments to get its mind off heavier matters"[2] would not likely criticize severely either the plays or the players.

Few plays of lasting renown were produced by these local companies. Serious plays like *Damon and Pythias, Re-*

94

becca of Sunnybrook Farm, Pygmalion and Galatea, or *The Lady of Lyons* were presented by high-school or college students. More typical of the offerings of the local "literary society" or "dramatic company" were such plays as *Hans Von Smash, the Blundering Dutchman; Tatters, or, The Pet of Squatters Gulch; Grace, the Poorhouse Girl;* and *The Sparkling Wine Cup.* The last play, and another favorite, *The Social Glass, or, Victims of the Bottle,* depicted the horrors of drink. Following the production of a play of this genre, one minister praised it, "sanctioning the play, and pointed out the moral it has shown, 'look not upon the wine when it is red.'"[3] Thus assured of the propriety of the production, the audience joined the pastor in congratulating all concerned upon its success.

Another temperance play that surely did not require an explanation of its message was *Arthur Eustace; or, A Mother's Love.* It depicted the downfall of a young man after he was introduced to the "Red Demon" and the mother's love that ultimately reclaimed her erring son. When the play was presented at the GAR Hall in Houston, "there were tears in the eyes of many in the audience when Arthur has disappeared in the prison cell and his devoted mother assured him that despite his sin she loved him still."[4] Comic relief in *Arthur Eustace* was provided by three national types, Hans Hurst, Patrick Flannigan and "The Chinaman," broadly played with heavy stage accents. Such caricaturing seldom failed to draw guffaws from an Ozarks audience, and dialogues, recitations, and comic songs, as well as plays, made liberal use of black, Irish, German (or "Dutchman," as the German was commonly called in the Ozarks), and Chinese stereotypes.

In this tradition was the comic play *Hans von Smash, the Blundering Dutchman,* produced at Ava (Douglas County) on Christmas night, 1891, for the benefit of the Ava Cornet Band. A synopsis of the play indicates the stress placed on the comic accent of the "Dutchman:"

95

Act I: Hans hires both his hands and takes good advice. Katie objects to his "furrin' ways."

Act II: Hans has experience with the chores and doesn't like them. He waits for "dem tramps" and they finally come. He opens the oyster can "mit happiness." ... Hans tries "dot friendship" and concludes "he likes dem." Batch apologizes for Hans and admits his right to the name of von Smash.[5]

"Pete the colored boy" delighted the reviewer of a drama entitled *The Deacon,* which was "very pathetic" but was relieved by the ridiculous antics of Pete: "For instance when Pete ... made love to the unsuspecting Miss Amelia who thought she was being wooed and won by the deacon, and again when the deacon made love to Pete, he mistaking the pesky Pete for his much adored Amelia."[6]

When home talent produced the "original border drama" *Tatters, or, The Pet of Squatters Gulch* ("proceeds to go for the benefit of taking the post out of the opera house"), the editor of the *Christian County Republican* erupted in an extravaganza of praise: ". . . the best thing seen here in five years. It was immense, . . . it was superb. Much better than any travelling troupe that has visited our opera house." "Mose, the black servant" (and, as usual, the comic relief), was singled out for praise as a "honey, a lulu, a peach."* A summary of the complicated plot shows that it was a "melodrammer" of melodramas:

> John Merston, a wealthy man went to Nevada many years before the time of the play, bought up several thousand acres of land, married a beautiful Indian woman and then deserted her. On his return to Nevada he found the lands valuable and many squat-

---

*Black characters were invariably played by white actors and actresses in blackface. The relatively few blacks in the Ozarks had their own entertainments, which were often well attended by whites, but they did not appear with whites on stage in a play.

ters living upon them. He also discovers that one of
the squatters, Robert Ferris, is the son of himself and
his Indian wife who has recently died after a life of toil
and privation. While Marston and Ferris are talking,
Marston falls over a high mountain precipice and is
supposed to have been killed. Squatters appear on the
scene and accuse Ferris of having murdered the old
man by pushing him over the cliff and they try to
lynch him but are foiled by the sheriff and Major Timberlake, landlord of Squatters Gulch and his daughter,
Titania, or Tatters. Phil Dolan, a leader of the lynchers
is in love with Tatters but Tatters loves Ferris. She
conceals him in the house assisted by Mose, the colored
servant. Ferris escapes but returns to declare his love
for Tatters and to take her with him. The lynchers
capture them in the woods but Marston reappears and
the lynching is again prevented. Phil Nolan gets his
deserts. Marston and his son are reconciled and Major
Timberlake rejoices with the lovers.[7]

The praise heaped upon *Tatters* was not atypical of the
accolades of hometown reviewers throughout the Ozarks.
"We don't mean to flatter our home talent," wrote a Walnut Grove playgoer, "but we really imagined that we were
in a first class opera in Chicago, Cincinnati, or New York."[8]
The editor of the *Cassville Republican* quoted unbiased authority in approval of the Seligman Dramatic Company's
rendition of the dramas *Wild Mab* and *The Reward of Crime:*
"They play well and would do credit to a much larger
town. Passing opera troupes of a national reputation fail
to hold the interest and in fact do no better than your
home players did tonight," remarked Charles F. Lawson,
a Ft. Smith travelling man.[9]

The heroine of the Ash Grove production of *Broken
Links* in 1895 was "good in posture and facial work,"[10] and
the tragic performance of Georgia King in the title role
of *Sweet Briar* "did great justice to herself and the company.
. . . One would imagine that she and the deacon were the-

atrical people of more than ordinary ability."[11] The characters of the Ash Grove College version of *Damon and Pythias* "were in every instance well sustained and some played their role with superb effort, even with brilliancy."[12]

Few companies failed to add musical embellishments to their plays. "Between each act and all through the play there will be specialties, songs and good music," read the ad for the five-act play *Carl Johnson the Woodman*.[13] The Ash Grove Silver Cornet Band accompanied the Ash Grove Literary Society to Walnut Grove to open Mr. Toalson's new opera house in 1892,[14] and the Hungry Hollow String Band shared the musical program with the Cassville Brass Band when the Cassville Dramatists presented *Down by the Sea*. The brass band appeared in full uniform and rendered new selections, and "their appearance upon the stage was as pleasing to the eye as their music was to the ear."[15] Vocal, cornet, and piano solos augmented other productions.

A dramatic production that achieved even moderate success in its home stand was virtually assured of limited touring prospects. The Purdy Dramatic Club, a particularly active organization, frequently took its shows to neighboring towns. The play *Tony the Convict* was revived, by popular request, two years after its first production and presented in Cassville. The audience was too large for the hall, and it was repeated the following night and later played in Monett, Exeter, and Verona.[16]

An outstanding troupe was assembled in Barry County under the name Exeter Dramatic Company. In the early months of 1895, under the guidance of a Mr. and Mrs. Harper (Mrs. Harper was also the leading actress), the troupe played Exeter, Purdy, Cassville, and Little Rock, Arkansas, with a repertory of six plays, *Chick the Mountain Waif; Gyp the Heiress, or, The Dead Witness; The Old Maid's Triumph; Kathleen Mavourneen; Divorced;* and *The Little Duchess*. The company usually played one- or two-night stands, playing the same bill in each town. An exception was *Kathleen Mavourneen,* which played in Cassville for an entire week.

Mrs. Harper was supported by a group of Cassville "local talent" for one play and by Exeter people for the next. Reviews were very favorable, and attendance was high. The *Cassville Republican* of February 28, 1895, carried the notice of the play *The Little Duchess* and warned, "This will probably be the last opportunity of witnessing the happy performances of the versatile actors, Mr. and Mrs. Harper, and it should not be missed." The following week notice appeared that the play had been presented, after which the Harpers disappeared from the pages of the paper.[17]

Three other dramatic companies were organized in 1895 and 1896, perhaps inspired by the successful example of the Harpers. They, unfortunately, did not fare so well. One of these organizations apparently failed to progress beyond the selection of an impressive name. The founders of "Fortner's and Youngblood's World Renowned and Famous Shows" rented an empty tenant house on a conveniently situated farm to practice their repertoire of "trapeze, music, song and dance, wire walking, sleight of hand and legerdemain, in fact many other things too numerous to mention." It is sad that no more was reported about their plans for a worldwide tour.[18]

At about the same time that the promoters of "Fortner's and Youngblood's World Renowned and Famous Shows" were dreaming of fame and fortune, a Douglas County organization, the "E. A. Danns Show," was indeed on its way, though not destined to go far. The troupe's initial performance at Bradleyville was so well received, and the twenty-dollar box office so inspiring, that it proceeded to Taney City in high spirits. There fame proved too unaccustomed a burden to bear gracefully:

During the first act a dispute arose among the performers just the same as city actors and the curtain arose on a scene like unto a saloon row or the Donnebrook Fair with everyone trying to hit someone's head. This the audience enjoyed highly for a while as they supposed it was part of the performance, and that the boys were giving an exhibition of a saloon row but then the boys began to use brass knucks and draw their old musty revolvers and genuine bloody blood began to flow.

The audience hastily evacuated the building, and when the fracas was stilled, the company disbanded and the members quietly returned to their homes.[19]

The third of this trio of ill-starred theatrical lights achieved its modicum of fame as the "Apollin Minstrels." The company consisted of five boys, who allowed themselves several rehearsals in West Plains before starting out to play one-night stands in the small towns of southern Missouri and northern Arkansas. The Johnson Opera House in West Plains lent them several hundred tickets (mostly complimentary ones), and bills were printed, advertising their coming:

THE APOLLIN MINSTRELS
A company composed of artists well known
to the profession
THE FINEST MUSICAL AND MINSTREL SHOW
EVER PUT ON THE STAGE
Admission 25 and 15 cents

People who carry guns please keep
outside the guy wires

From this point the story of the "Apollin Minstrels" can best be told in the words of the hometown correspondent:

On a bright morning and in a covered wagon they left. Attie being their first stand. Here a small sized

100

crowd greeted them and they next tried Thomasville. At this place Joe Allen busted the head of his banjo and the boys had to lay off a day waiting the arrival of another banjo head. At Moody enough money was taken in to pay for the banjo repairs which safely arrived from West Plans and was hailed with joy. At Viola the box office receipts amounted to a dollar ten cents which paid the bills for horse feed. The next stop was at Bakersville and a full house greeted the minstrels when the curtain went up. While Onzello Dixon was singing "The Deacon's Dilemma," a big mosquito from Bennett's Bayou who had a free pass to the show took a reserved seat on the very point of Onzello's nasal organ. The bouncer of the show had to be called to fire the intruder out, which was done after a very lively scrap. At Gainesville, the boys got tired of cooking their own grub and put up at a tip top hotel. Here they played to empty chairs and when they started to leave the town next morning they found their baggage had been attached for board, and the Apollin's would have come to grief had it not been for the timely aid of several of the Harlan boys. From there they jumped down to Arkansas where Onzello sang a new song expecially composed for the occasion. As to their trip into Arkansas, not a word could be learned, but from the looks of the boys we suppose that the people of Arkansas made it very interesting for them, and they were very thankful to have escaped with their lives.[20]

While this account of the adventures of the Apollin Minstrels may tend to the apocryphal, the imaginative correspondent probably based his report on fact, however far he may have strayed in the telling. Perhaps he was a frustrated playwright, and his frustration exhibited itself in envy as he described the adventures of would-be actors who almost found a release for themselves. When Ozarkers could not enjoy a drama enacted on the stage, they created their own drama from whatever materials were at hand. Hence the Ozarkers' love of the tall story (*We Always Lie To Strangers* is the title Vance Randolph chose for a book about the Ozarks), which has become almost a trademark (or hallmark) of the mountaineer. The plays most enjoyed by the hill folk were not serious drama but exciting melodrama and hilarious farce performed for them by their friends and neighbors. Knowing this, and recognizing the delight to be taken in stretching the truth a bit, we find it easy to accept the pronouncement of an Ozark sage: "A little nonsense is relished by the wisest of men."[21]

Stultz pie supper has come and gone and oh we had a glorious time and we think everyone present enjoyed themselves at least they appears to. About twenty pies were sold I think and the proceeds amount to five dollars.

—HOUSTON HERALD, DECEMBER 12, 1901

Box and Pie Suppers

A N Ozark box or pie supper presented to the uninitiated a deceptively simple format. When a community wanted to raise money for some worthy cause, they often decided to do so by means of a box or a pie supper. The women and girls baked pies or prepared box lunches, and every-one gathered at the local schoolhouse, where the pies or boxes were auctioned off to the men and boys present. Purchase of a pie or box brought with it the companion-ship of the girl who had prepared the item. The proceeds of the evening went to the chosen cause.

Such a factual and skeletal definition is not incorrect, for it describes the supper's surface activities accurately enough. The real reason for the popularity of this event, however, is not readily discernible to the casual observer. A successful box or pie supper is built on ritual, and it is only with an understanding of the subtleties of the rites involved that such an event can be truly appreciated or even understood. To auction or sell a donated item for a charitable purpose is merely a practical and oft-utilized means of raising money. A pie supper was so much more than this. It was entertainment, fellowship, and participa-tion. It was courtship and a sporting event, charity and

103

food, all melded delightfully in a single ceremony that Ozarkers of all ages loved. Older farmers discussed crops and settled school-board business; children ran and laughed and played together, reveling in the infrequent companionship; young folks carried on discreet courtships; and the older women visited and traded recipes and gossiped.

The ritual began in the homes where the pies or boxes were being readied. Custom dictated that their owners must remain anonymous until the higher bidder came forward to claim his due.* A girl who was of courting age but was not seriously attached to any boy, would go to great lengths to make sure that her offering could not be identified prematurely. The sporting blood of the boys of the community was naturally aroused by such a challenge, and the order of business for the day of the "supper" was to associate boxes and pies with their owners before the auction. The first arrivals at a pie supper, therefore, were the young men, who lined the path to the schoolhouse door trying to sneak a glimpse of the tribute being borne inside.

As a measure of the counterattack, girls often traded pies or boxes among each other before taking them inside, or gave them to an older neighbor woman to carry. Another diversionary tactic allowed the girl to let slip a false hint about the way her package would be decorated. The young man who bid high on such an item often discovered after he paid for it that the person with whom he was to share his purchase, and the remainder of the evening, was not the girl he had thought but was an older married woman or spinster of the community. The discomfited bidder could do no more than grin sheepishly at the jibes

---

*The pie or box was usually identified only by a number. At one box supper in Taney County, however, the boxes were sold by "shadows" of the girls. Patrons were urged to "study . . . your girl's shape for the next week and back your judgment with your money." *Taney County Republican*, February 27, 1896.

and torments of his friends, for this was part of the game, one element of the ritual.

While ostensibly taking great pains to conceal it from view, a girl might secretly tell her best beau how her box or pie was wrapped so that he could recognize it and bid for it. Such bidding had to be done cautiously, so as not to give away the secret. This technique involved the boy's setting up a smoke screen by entering the bidding on several other offerings as well, making sure not to bid too high on them. When *the* box or pie was presented, the greatest casualness and indifference had to be affected. If a trustworthy friend was present, he might carry out the bidding by proxy. These elaborate machinations were necessary, for in a favorite rite of the suppers a group of boys would decide to gang up on one of their number and outbid him for his girl's pie or box. The beau's status and honor in the sight of the community and his girl depended on his topping the bidding of all competitors. To accomplish this, it was sometimes necessary for him to gather his own coalition and pool cash resources. No matter which side won the bidding, such attention was highly gratifying to the owner of the box or pie, and the delighted girl usually ended the evening by sharing her food with all her victorious admirers.

Fancy wrappings and decorations for all the boxes and pies were in order, and a table loaded with such items provided a colorful stage setting for the evening, delighting the eye and whetting anticipation for the events to come. In some communities "married" pies and boxes were kept separate from the others. These were disposed of rapidly to the proper husbands so that the more interesting business of the "unmarried" items could be attacked. At most suppers however, the contributions of all children, young people, and married women alike were piled together, and the interest and suspense was thereby intensified. Not all the identities were kept secret. It was a point of great pride

An auctioneer calling for bids on a fancy box at an early-day pie supper in Mincy, Taney County (note the deer antlers spelling "Mincy"). Photograph by Townsend Godsey.

with many a matron that the bidding for her pie was lively because of her reputation as a pie maker.* Some suppers featured "bachelor pies." Those who bought bachelor pies received only the pies—no feminine companionship accompanied them. Naturally, the selling price was usually low and was a real bargain for the individual who was interested primarily in food.

The focal point of attention, once the bidding got under way, was the auctioneer. His task was not merely to hold up the items for sale and accept bids but to amuse the audience with witty remarks, make jibes at the bidders, and competently field the insults that would inevitably be returned. All this he did while encouraging the bidding as high as possible. A good auctioneer contributed greatly to the financial success as well as to the entertainment value of a supper, and his services—gratis, of course— were in demand.

A regular feature of most suppers, and the highlight of the evening, was the contest for the "beauty cake." A donated cake was placed on display, to be awarded to the girl whose admirers contributed the most votes, at a penny a vote. Joe McKinley explained the procedure of this event:

> They would run Miss Black and Miss Brown. Someone would be interested in Miss Black, and they'd come around: "How much do you want to vote at a cent a vote?" Then they went up the checkers who sat there, and if you had four dollars this trip for Miss Black, they marked her up four hundred votes. Then they put a time limit on it, and when that was up, why that was that.

Occasionally such a contest would realize fifteen to twenty-five dollars, though smaller figures were more common.

*May Kennedy McCord recalled that a lemon pie was considered "stylish," and was always in great demand, probably because of the scarcity of fresh lemons in the Ozarks in early days.

Once in a while, when circumstances were right, a record-breaking vote would net the cause an unusually large amount, as in Oak Grove (Douglas County) in 1907, when the beauty cake sold for $74.35.[1] Such enthusiasm usually resulted when an adventuresome group of boys (probably from another community) became determined that their contestant must win. Sam Miller told about a group who went to every pie supper within an area of fifteen to twenty miles, putting up the same girl for the beauty cake: ". . . it finally got to be almost a traveling pool. They'd take her, and all the people on Wolf Creek would go along, and they'd back her up. They'd save their money from one month to the next. . . . these old boys at Wolf Creek, they would spend every dollar they had and arrange to borrow."

Spiritual kin to the Wolf Creek boys were the Stony Point boys, of Douglas County: "The voting was exciting for a few minutes before the Stony Point boys went bankrupt but they had one in their crowd who had ready wit to meet such imergencies [sic] and brought in his saddle and laid it at the feet of the collectors. This brought loud applaus [sic] from all the house."[2] The effect of this dramatic gesture was shortly nullified when it was discovered that the saddle was "borrowed," and so the opposition candidate carried off the cake.

At West Plains one young man, "in order to have his contestant win the prize, burst his bank account, borrowed all he could from his friends, pawned his watch and chain, and being still a little short, he pawned his overcoat also and this took the cake."[3] Joe McKinley told of one pie supper at which a group of boys picked the homeliest girl in the community, entered her name, and won the contest. Other contests were held for prizes—a jar of pickles for the most lovesick couple, a cake of soap for the man with the dirtiest feet—but none approached in popularity the beauty cake for the prettiest girl.

Most box or pie suppers were held for the benefit of the local school. Emmett Yoeman explained:

> Of course tax rates were low. Taxes furnished nothing in an incidental fund for supplying needs there for maps or globes or library or the extra things that were necessary for good teaching. So the more progressive teachers would have pie suppers or box suppers, with a program, of course, to try to raise some money for books or extras that would help them in their work.

The $15.65 collected at a Stone County pie social was expected to go a long way. It was earmarked for "a book case for the school library, reading chart for beginners, small globe for geography class, one set of arithmetical blocks, one large wall map and remainder to be invested in books for the school library."[4] Elk Creek, in Texas County, held a "belfry box supper" to raise money for a new belfry for the schoolhouse.[5]

Besides school benefits, box and pie suppers were held to help pay the minister's salary, to maintain the graveyard, to help pay for a church organ, to purchase band uniforms, or to help support the high school football club. Benefit suppers for individuals were not uncommon. The proceeds from a pie supper at the Raymondville Church (Texas County) went to "Widow Graves, children and blind mother who are very poor and helpless."[6] Neighbors "got up" a pie supper for Aunt Man Bush when her first husband died and made forty-six dollars for her.

The proceeds from a box or pie supper rarely ran more than fifteen dollars, and more commonly eight or nine dollars. High receipts were usually the results of the sale of the beauty cake, for boxes and pies sold for about twenty to twenty-five cents each, though spirited bidding would run an occasional item to over a dollar. Most "suppers" had no planned entertainment beyond that furnished by the event itself. Occasionally however, home-talent musicians performed, or declamations, recitations, and dialogues were presented. A play entitled *The Black Heifer*, interspersed with songs and instrumental music, regaled those attending the box supper at the Homeland Church in Howell County in 1910.[7] The two hundred people who attended the box supper at the Scott schoolhouse in 1907 saw three plays and were entertained by a "phonographist" between acts.[8]

Conduct at pie and box suppers was surprisingly good. An occasional flare-up of tempers, usually ignited by the competition for the beauty cake, was viewed with tolerance by the elders of the community, for such excitement was usually accompanied by a healthy flow of cash into the till. Many visitors to these affairs were from outside the local school district, and their money was regarded with favor by the taxpayers of the community. Mrs. Eva Dunlap recalled:

> Now these old Dads, they didn't buy any even if they knew their wife had a pie there. This was from their district and they didn't want to put their money in it. Let it come from the outsiders and young folks who didn't pay taxes, and make up what they could and if they wanted to put in any more afterward they put it in but they wasn't a-gonna spend.

Box and pie suppers provided audience-participation entertainment in a ritualistic form that may have been puzzling and obscure to the uninformed but was immensely satisfying to the participants who understood its subtleties.

Everybody is waiting for everybody else to hold the first picnic at the big spring. The first crowd gathers up all the chiggers.

—LAWRENCE CHIEFTAIN, JUNE 7, 1894

Picnics

EARLY-DAY Ozarkers had an enormous capacity for enjoying people. Forced by circumstances of geography to live a more or less isolated existence, they welcomed any opportunity to meet socially with friends and neighbors.

The members of the Howell County community known by the delightful name Peace Valley, held a picnic each year for themselves and neighbors of the nearby countryside. By the last of August most of the crops had been laid by, and the farmers felt free to relax after the drive of the earlier planting and harvesting season. A correspondent who wrote to the *West Plains Journal* on August 26, 1897, seemed to be in a relaxed, congenial mood as he discussed the Peace Valley picnic of that year:

> The people count on going to this picnic each year just as regularly as a young man goes to see his best and sweetest girl. One always has a pleasant time there. The people are the most hospitable in the world, bountiful and well prepared provisions are provided for all, and people are jolly and happy.
>
> They give the day to pleasure, visiting one another, speech making, swinging, dancing and helping the boys out who have lemonade and refreshment

111

A formally dressed picnic group posed in a rustic setting in Christian County. Courtesy State Historical Society of Missouri.

stands. . . . The young people are as well behaved, cultured and refined as any community can boast of and we heard a number of compliments to the beauty of the young ladies present.

There seemed to be only three requisites for an Ozark picnic: a good spring, plenty of shade, and the desire. The first two were to be found in abundance, and the last was seldom wanting. The result was a plentiful occurrence of one of the Ozarks' favorite warm-weather entertainments.

Any location that was convenient to the people and had water for both horses and human beings might serve as a picnic spot. Galena's favorite spot was an island in the James River. In 1903 the big new railroad tunnel at Reeds Spring was a source of wonder, and many picnics were held there.[1] Crossroads stores played host to many celebrations in the Ozarks, particularly in Douglas and Ozark counties, where towns and villages were few and far between. Other favored locations were river fords and shaded groves. Most towns had established picnic grounds or parks where gatherings of all kinds were held.

Although a picnic could be held for no real reason whatever except sociability (as was the Peace Valley affair), certain special occasions lent themselves particularly to this form of celebration. The Fourth of July, Decoration Day, veterans' encampments, and old-settlers' reunions were the most notable of these affairs.

FOURTH OF JULY PICNICS

A single issue of the *Taney County Republican* for June 30, 1904 carried notices of Fourth of July picnics at Antioch, Branson, Brown Branch, Bluff, Cedar Valley, Dit, Groom, Kirbyville, Protem, Taneyville, and Walnut Shade. There was seldom want for a July Fourth celebration in the Ozarks. If one wished, he could take in more than one, as did an editor in 1901, who rode forty miles to attend two picnics

113

Picnickers refreshing themselves under an Ozark brush arbor.
Courtesy State Historical Society of Missouri.

and a fish fry.[2] These "jubilations" ranged from simple basket dinners and sociability to elaborate all-day programs with parades, orations, and fireworks. Although attendance at the latter variety was usually large (6,000 at Ash Grove in 1887[3] and not uncommonly 1,200 to 1,500 at picnics), small-town merchants often needed prodding and pushing each year to organize and sponsor them. The editor of the local paper seldom failed to take upon himself the task of gadfly, and for five to eight weeks preceding the Fourth he annoyed, coaxed, begged, threatened, and bragged the town into action.

A typical opening shot is this one from Gainesville, whose comma-ridden copy discloses wry editorial humor: "The Anervesary of the independence of our country, comes on the fourth of July, as usual, this year, no changes having been made. The question now, is, will Gainesville enthuse? If so, it is a matter that should come under discussion, by our people at once."[4]

The urging usually succeeded, the celebration was a glorious success, and post-picnic issues of the paper were devoted to laudatory comments about the fine cooperation, with only gentle chiding for the few mossbacks who dragged their feet. Occasionally the plans for the celebration fell through, and bitter editorials appeared:

> Ash Grove was not at home on the Fourth. Ash Grove did not celebrate and so it went to see the neighbors who did. Our town spread itself out over considerable territory that day. It went to Springfield and Fort Scott and Golden City and Walnut Grove and Dadeville and Cave Springs and Phoenix and Comet and Haven and several other places. It was quite promiscuous. It spent as much for railroad fare as two celebrations would cost and didn't have half the fun it could have had at a picnic at home.[5]

Houston tried for several years to have a celebration on the Fourth but seemed not to be able to achieve it. The editor was magnanimous in his explanation: "Houston has held the courts, conventions, institutes and many other similar gatherings, and it was considered but fair that this particular holiday should be left to the other parts of the country."[6]

Fourth of July picnics were well publicized not only by posters and flyers but by newspaper articles and occasionally full-page illustrated advertisements. When the program was listed, much stress was placed on the promise that "the events will be carried out as advertised." Editors and correspondents had no sympathy with the management of an entertainment that did not fulfill its promises. Ozarkers did not mind simple fare, but they expected faith to be kept with them. An indignant editorial writer expressed his disgust with a picnic thus:

> In fact there was no program and the whole thing went flat. We have been party for the last time in booming a picnic or reunion that does not give some substantial assurance that the program is going to be carried out. It is simply wrong to humbug the people as those who attended the picnic here yesterday and hereafter we will not be a party to any such business.[7]

More typical, however, was the satisfaction and pride evinced in Truman Powell's letter to the *Stone County Oracle* of July 7, 1904. "Not a single failure occurred," he glowed. "The program was carried through all complete."

Most accounts of picnics reported perfect order throughout the day. Powell attributed the fine order at Marmaros to the "good substantial citizens" who made up the large crowd. "There was no whiskey on the ground nor was a drunken man to be seen,"[8] and in almost every instance of disturbance at such affairs "Johnny Corn" was in some way involved.

The combination of the twin evils, liquor and dancing, produced particularly explosive situations. Knifings on

116

the dance platform were not uncommon, and considerable effort was expended in many communities to prohibit dancing platforms from picnic grounds. Upon hearing that the plans for the celebration at Forsyth would forbid a dance platform, an editor exulted breathlessly:

> It is high time that our people would dispense with the ancient rites of barbarism and Indian corn dances and red liquor feats and pop gun and butcher knife accompaniments to take up the more elevating and educative methods of civilization in perpetuating the memories of our forefathers in securing to us a day that we may all reach up and pluck a feather from the Old Eagle's tail in a way that will not be humiliating to his venerable Birdship.[9]

A Butterfield correspondent gloated: "There was a dancing platform erected but thanks to the good judgment of the girls there was not enough inclined that way to make a set so the poor lonely platform is still waiting for occupants."[10] Other picnic disturbances included reckless driving in buggies. In one such instance a small boy was injured at Mount Vernon.[11] The cupidity of an official created considerable interest at a Fourth picnic at Aix. The crowd was leaving the grounds after the picnic when a "squire" (justice of the peace) noticed a young man, slightly drunk, carrying the barrel and stock of an old pistol but without cylinder or hammer. According to the correspondent, "The Squire had had no case since he had been made an officer on the sixth of November and he thought it was about time," so he, with the aid of a constable and another man, arrested the boy. The squire asked the boy how much money he had, found that the amount was one dollar, and promptly assessed the fine for that sum. He then divided the money with the other men and turned the boy loose. By this time an indignant crowd had gathered and witnessed the proceedings. There were angry mutterings,

117

and the squire was given ten minutes to return the dollar to the boy. That he did, posthaste.[12]

The larger picnics usually featured a parade that initiated the day's festivities. Country people rose early to do their morning chores and be in town to see the spectacle, arriving sometimes by sunrise. In 1895 the Cassville-Exeter band met the nine o'clock train, which was loaded with visitors, and serenaded their arrival.[13] A Fourth of July parade was made up of whatever marching and riding elements could be persuaded to participate. A band, of course, was nearly indispensable. Merchants occasionally sponsored floats displaying their names, and bicyclists and young men and women on horseback added to the color and excitement of the event.

In 1898 the Spanish-American War inspired added patriotic themes, such as Uncle Sam befriending a Cuban and a marching platoon in Admiral Dewey uniforms.[14] George and Martha Washington, Uncle Sam and the thirteen colonies, and a Calathumpian band ("funmakers") marched at Gainesville in 1891.[15] In 1893 a favorite float of the Ash Grove editor was one drawn by four large horses. It represented "Columbus before the court of Spain asking for help to aid him on his voyage." The costumes were described only as "flawless," but the reporter's appreciative gaze dwelt on Miss Nellie Dixon, who, "with her dark hair and eyes, eyes which looked down into so many souls with a kindly light to always find a responsive chord, fit the part of Isabella while by her side was Mrs. S. O. Weir, every inch a Ferdinand."[16] The next float showed Indian life as Columbus found it and carried a wigwam and townspeople made up as Indians.[17] Few were the parades that did not have floats depicting the states of the Union and Miss Liberty: "The 44 states were represented by that number of probably the best looking corpse of young ladies that ever performed that delightful task."[18] The same Ash Grove editor described the Goddess of Liberty, repre-

118

sented by a young lady who "has been lavishly bestowed upon by the hand of nature, whose every feature rendered her the most suitable of all for this distinguished honor as it required an elegant, portly, lovely, and dignified lady, all of which Miss Nettie Nickle is personified."[19]

The grand theme of Fourth of July celebrations in the Ozarks was, of course, patriotism. "The American Eagle will spread his wings and soar aloft over the summits of the Ozarks," wrote Truman Powell. "Patriotism should be taught to the rising generation and the Fourth of July is the time to do it."[20] To accomplish this end, every celebration worth its name included someone reading the Declaration of Independence as well as a star-spangled assortment of orators.

It was no uncommon honor to be chosen to read the Declaration of Independence from the platform. The audience listened to the words of that document with a reverence approaching that attendant upon the hearing of Scripture, and the prime requisite of such a performer was that he or she be able to speak in a loud, clear voice that could be understood by all within earshot. The noise of the circle swing and the barking of the lemonade vendors often rendered this no mean feat. Newspaper accounts reflect the emphasis given to clarity and volume: "Notwithstanding the reader had to face a strong wind, her reading was good and was heard by all who could get seats near the music stand."[21] The Ava correspondent admired the reader who "read with a resonant tone making plain every word that fell from his lips,"[22] and at Ash Grove the people marveled at a five-year-old boy who recited the Declaration of Independence "from first to last without a prompting."[23]

The Cassville celebration of 1895 was a decided success in every respect but one. "Through the neglect or contrariness of the committee who had the management" there was no featured orator. After a few extemporaneous speeches (a pitiful substitute) the program fizzled.[24] The orator was a traditional and honored feature of the Fourth of July celebration, and one might as well dispense with the basket dinner as with the speaker's contributions to "spreading the eagle." One year the folk at Ava found their orator to their liking; he "spoke with such an oratorical tone of voice that every word pierced the hearts of his listeners making plain every thought."[25] The qualities desired in an orator were given in a description of an Ash Grove speaker: "Mr. Robertson is as eloquent as he is loud, as logical as he is lengthy, and as entertaining as he is interesting."[26]

Normally the Fourth of July speaker dwelt upon national, historical, and patriotic themes with the view of capturing his audience emotionally. The orator at Ellis Prairie in 1901 evidently succeeded in this: "His heart seemed to leap when he spoke of the signers of the Declaration of Independence risking their lives and putting their signatures to that glorious bill of rights that had been handed down from generation to generation known as the Declaration of Independence."[27]

Another Ash Grove orator appealed to his women listeners by emphasizing their right to participate in the civil duties of government. A reporter characterized his speech as being "full of that spirit of American patriotism which rises above political machines and partisan rancor. . . . His oration was broad, liberal, eloquent."[28] An Ash Grovian, G. M. Carter, wrote to the editor of the *Commonwealth* about his speechmaking experience at Aldrich. His letter gives some insight into contemporary oratorical form, as well as his personal style: "The speeches were characteristic of the true American Orator sound principles good logic rhetorical bursts of stelliform curvalinear beauty, except mine, which tried to stretch the eagle's wings so as to split the air (as the darky said) with a big 'acclimation' of American 'declipendence.'"[29]

While the traditional long flag-waving speech was very common, there were those who preferred another kind. A writer in the *Marshfield Chronicle* described a speech that was a surprising departure from the contemporary norm:

> Mr. Wisby entirely forsook the old beaten paths of spread eagle oratory and laid out thoughts that are contiguous with our every day life; problems that are confronting us every day that we live; certain features of legislative government that need remedial actions regardless of partisan politics. It was a masterly effort delivered with the ease that bespeaks not only careful speech training but natural ability as well.[30]

This form of public speaking—a logical presentation of contemporary questions—was again favorably noted, this time at Ash Grove:

> Mr. Godwin's speech was hardly of the regular Fourth of July brand such as we are accustomed to hearing. The speaker instead of indulging in an extravaganza about our country or frantically waving battle flags at the balance of the known world, talked of economic questions and the enforcement of law. Mr. Godwin is a man of mature thought. His speech was well composed and delivered.[31]

Before, during, and after the oration of the day most picnics offered other attractions. At smaller affairs these might be nothing more than good conversation with neighbors not often seen and abundant quantities of cold spring water or barrels of ice water hauled in for the occasion. At large picnics families gathered at noon to eat the basket dinners they had brought along. At the smaller picnics families placed their food on a large table, and the resulting feast was shared by everyone. Some picnics proudly featured a barbecue, as at Alton in 1896, when two beeves, eleven sheep, and six hogs were killed and barbecued.[32]

The circle swings mentioned above, similar to modern merry-go-rounds, were regular features at the picnics. The first swings were propelled by manpower or horsepower; later models were turned by a puffing steam engine.

Lemonade stands did a thriving business and were usually the only concessions on the grounds. The celebration at Ash Grove in 1892 had "a gambling outfit on the ground from Springfield, who relieved a few of our boys and one or two old men of their hardearned dollars," but the game was stopped, and the gamblers running the game ordered to leave the grounds.[33] Mostly the diversions consisted of games, races, and contests with prizes. Greased-pole climbing; foot races of every kind, including a fat men's race; bicycle races, and egg races were standard items on the program. Another favorite, the tug of war, was given added interest at Ava in 1910 when twelve country women were pitted against twelve town ladies.[34] The best jumper in three standing jumps won a two-dollar hat at Ash Grove in 1893.[35] There were prizes for both men and women in the hitching and harnessing contest at Ozark in 1901.[36]

The special attraction of the afternoon's entertainment was often a baseball game. If the local boys had been having a good season, the fans watched the contest and cheered their favorites with a loyalty befitting the patriotism of the day. One can imagine, however, that both spectators and players were a bit surfeited with victory at Ava when, in 1892, the Ava boys swamped visiting Bryant Creek, 63 to 5.[37]

Outside attractions were sometimes brought in for the day. People gaped and thrilled to balloon ascensions and parachute jumps, which were usually performed by a "professor" or a "madame." At Red Oak the correspondent wrote to the *Lawrence Chieftain* on June 27, 1889, about another popular attraction: "Negotiations are in progress for one of Professor Edison's talking machines (phonograph) which if successful will be an interesting attraction." The popular shoot-the-chute ride at Ozark was the cause of a tragedy in 1902. The ride consisted of a boat that hurtled down a slide and splashed into the river below. After operating without incident all day, the boat hit the water in such a way as to throw all thirteen occupants into the river, and an eighteen-year-old boy was drowned.[38]

The attendance prizes awarded at Ash Grove in 1893 were typical of the bounty dispensed by the management at picnics throughout the Ozarks to those who met certain stipulated qualifications:

1. Oldest Married Couple. $5.00.
2. Largest Family. $5.00.
3. Oldest Man. Rocking chair worth $2.50.
4. Oldest Woman. Fine dress pattern worth $4.00.
5. Ugliest Man. Half dozen photographs of himself.
6. Prettiest Lady. Half dozen photographs of herself.
7. Laziest Man. Umbrella worth $1.00.
8. Prettiest Baby under one year. $2.50 gold piece.
9. Largest Married Couple. $5.00.
10. Couple getting married on the stand. $5.00.
11. Farmer who raised the greatest amount of wheat in 1892. $2.50.
12. Farmer who raised the smallest amount of wheat in 1892. $1.50.[39]

Galena added prizes for bringing the largest number of people from the country in one wagon (the winner brought thirteen); to the man who had cast the most votes (15) for president in Stone County; to the woman having the largest number of children present (eight children, awarded twenty-five pounds of flour); and to the man with the largest feet (no measurement given).[40] Ozark rewarded the man with the baldest head, the man with the tallest family, the the man with the longest beard.[41]

After the orators had finished and the prizes had been awarded and darkness had fallen, there would be fireworks, but many of the country folks could not stay: "As the shadows lengthened in the grove, remembering the weary miles back home, the crowd started breaking up, tired but very happy. Some of their friends they met today they would not see again 'till next Fourth of July. This was 'the end of a perfect day,' and one long to be remembered."[42]

DECORATION DAY EXERCISES

No whit less patriotic than the Fourth of July celebrations were those commemorating Decoration (or Memorial) Day. The Civil War was still vivid in the minds of the people of the Ozarks, and while rancor had died down, May 30 was a day to stress harmony, friendship, respect for the dead (especially the soldier dead), and patriotism. An Ozark County correspondent indicated the importance of patriotism in these celebrations by capitalizing the word each time he used it. From him also came the information that "nearly every cemetery of any importance in the country held memorial exercises on Monday."[43] Sometimes the railroad offered half-rate fares to those attending Decoration Day exercises.[44]

A procession to the cemetery to distribute "flowers and warm tears" was a regular feature of the exercises, large or small. At Gainesville the lineup consisted of two visiting brass bands, veterans, sons of veterans, forty-four girls representing the different states, and citizens.[45] Ash Grove offered its visitors Springfield's military band and also its uniformed light guard ("They number over one hundred and took first prize in state competitive drill").[46] At Forsyth the procession was simpler. Following exercises in the church, "the assembly marched to the cemetery and decorated the graves of both Federal and Confederate soldiers alike. The singing and grand-march by the children was beautifully executed.[47] It was there also that the orator made an impressive speech "and left no barriers between the heroes of the North and the South."[48]

Ministers often spoke, and several shorter speeches, rather than one long one, were common. Recitations, songs, and instrumental numbers rounded out the platform program. "A part of the exercises worthy of mention was the decoration of the monument to the unknown by the forty-four girls representing the states each of whom recited an appropriate verse while bestowing the flowers."[49] Songs by choirs, soloists, and instrumental groups were of a patriotic and sentimental flavor. Some representative song titles are "Land of Freedom," "Cover Them Over," "Dixie," "The Faded Coat of Blue," "Marching Through Georgia," and "The Gray Beneath the Blue."

Levity was not tolerated at Decoration Day programs. The Ozarkers' great respect for the dead and for the old soldiers to whom, of all the living, the day belonged forbade frivolity in speech or song. Ozarkers dearly loved political speeches, but even they were out of bounds on this occasion. A political rumor was started against a young prosecuting attorney in Stone County that he had made a political speech at a Decoration Day exercise. He vehemently denied the charge and deeply resented the implication that he would even consider doing such a thing.

The opportunity of neighbors to assemble, to talk of the past, to eat together, and to be regaled with music, speech, and spectacle, while honoring the memory of loved ones and of soldier dead, made the Memorial Day exercises universally popular.

OLD SETTLERS' REUNIONS AND OLD SOLDIERS' ENCAMPMENTS

Old settlers' reunions were designed to provide a gathering place for old Ozarkers and an opportunity to honor them. The old soldiers' encampments purported to do the same thing for the former wearers of the Blue and the Gray. That these functions were fulfilled there is little doubt. That they were very popular is a fact. But it is also true that the old settlers and the old soldiers were far outnumbered by those who saw in these occasions a wonderful excuse for a good time. Of the 5,000 people who jammed Ash Grove for the Grand Army Encampment in 1892, only 366 were registered as old soldiers.[50]

Reunions usually lasted three days, occasionally four or more. Families traveling from a distance camped on or near the grounds in tents or covered wagons. The railroads usually reduced rates or offered excursion rates to the reunion sites. Attendance of 2,000 to 5,000 people was common, and good order was the general rule: "Kind feeling and good order prevailed from the beginning of the reunion on Wednesday morning till the end on Saturday night. There was only one attempt at a scrap and it did not amount to anything."[51]

The reunions were held in August and September, after the crops were laid by and the farmers could be away from home for a while. They were carefully planned and executed, and the extended period of the reunion or encampment provided not only time for visiting but good entertainment as well.

At the old settlers' reunion at Phelps, Lawrence County, in 1887 there were "five or six swings and fifteen to twenty lemonade stands on the grounds." The Forest Home band played concerts, there were several speeches, attendance prizes, a tightrope-walking exhibition, and a ball game.[52] Houston provided much the same in 1906 but added balloon ascensions, fireworks, a debate between two candidates for Congress, a greased-pig chase and greased-pole climbing, a fiddler's contest, and prizes to the fattest man (345 pounds) and woman (314 pounds).[53] Three years later nighttime crowds were thrilled by the electric lights that illuminated the grounds. "Two hundred and fifty dollars worth of free entertainment was provided by the powers that be," including a trapeze troupe from Kansas City, and Captain Bodardus, "a champion shot of the world."[54]

Everything that an entertainment-seeking Ozarker could desire was on hand at Clarkston, Lawrence County, in 1894: ". . . well dressed men, pretty girls, sweet babies, and abundance to eat."[55] This Mecca might have been foreseen four years before at Phelps by the gypsy woman who "predicted the fates of love sick swains at the rate of two dollars per predict."[56]

With so many things to do and see, old settlers did not have much time for speeches. Some speeches were essential, of course, but they were kept to a minimum, both in quantity and in length. One man who showed up to speak at Crane was disgusted by the lack of interest in speechmaking:

> I found that there were no facilities prepared for speaking, neither could I find any member of the committee. There was a platform erected at one end of the grounds which I supposed was placed there for the accommodation of speakers, but there were no seats provided for the listeners. The band was stationed at the other end of the grounds and speaking under the circumstances was entirely out of the question.[57]

The other two speakers scheduled for the occasion must have had advance warning of the odds against them. Neither showed up. One went to the county seat to attend to some political business; the other stayed home to boss his threshing outfit.[58]

Seven thousand people crowded into the little community of Phelps for the reunion in 1897, and besides hearing talks by a dozen old citizens, three politicians, and a preacher, they listened to a contest for the best original oration and the best original declamation, the winners each receiving a prize of five dollars. One might imagine that

This Grand Army of the Republic reunion took place at Bradleyville, Taney County, about 1905. Courtesy Douglas Mahnkey.

the interest was keen in the contest to pick the prettiest widow between the ages of twenty-five and forty. A total of $2,500 changed hands during the two days of the picnic, and everyone left happy: "The stand men all made money and the merry-go-round came out $80 ahead."[59]

The reunions of the old soldiers resembled the old settlers' reunions in most respects. The former affairs had a bit more martial air—campfires at night, military drills, patriotic declamations—and the old soldiers made their wishes known in regard to what they did and did not want. They did *not* want dancing in or around their encampment at Taneyville in 1898, and, much to the chagrin and disappointment of many young people, their wish prevailed. The editor of the *Taney County Republican* chided those who would go contrary to the veterans:

> The gathering is to be a reunion of the soldiers of both sides as in the Civil War and not a picnic. Every person besides the veterans who attend come as invited guests and will be welcome, but invited guests are supposed to conform to the rules established by the household and not set up rules for themselves. Many of the veterans will attend with their families if there is no dancing platform who will not come if one is maintained. Of course none will be on the grounds proper nor should there be one in Taneyville or anywhere near. Nor do we think the people of Taneyville can afford to allow one. The morals of that place have been paraded to the world and not unjustly either. Now when the old men who compose the veterans of the county vote to hold their reunion at that place on account of its morals as much as anything else can that people be first to be immoral?[60]

Five years later the scene had changed considerably. Perhaps the demise of many of the old veterans had weakened their collective voice. In any event, at the encampment of

1903 "there was no welcome address, no speeches, and those so constituted so as not to enjoy a big dance or a big drunk found themselves quite lonesome most of the time."[61]

After the turn of the century old soldiers' reunions were held sporadically. Only forty-seven veterans registered at Kirbyville in 1903 (forty Union, seven Confederate), and the affair was so slightly regarded that two threshing machines worked in the neighborhood during the entire time of the reunion.[62]

Entertainments at the old soldier's encampments were not usually as elaborate as those enjoyed by the old settlers. The veterans seemed to be satisfied with talks, music, and some dramatic entertainment. The Ava correspondent tells how it was in 1901:

> At night there was a very interesting programme carried out consisting of marshal music, singing by the choir and short talks was the veterans present. Uncle Sambo and Aunt Dinnah gave a very interesting sketch of negro life which was appreciated by those present. After which attorney John W. Murray delivered the address which was listened to with marked attention.[63]

A minstrel troupe from Arkansas entertained veterans at Cassville in 1897, but "the Modoc Indians got on a drunk at Seneca and failed to materialize." This disappointment was compensated for by a black orchestra from Fayetteville. Not only did they furnish fine music, but their "gentlemanly behavior and bearing was second to none." The editor added magnanimously, "Give the colored brother his due when he earns it."[64]

In 1894 the Cassville soldiers encamped to the music of four different bands and drum corps and viewed a parade led by forty-four young ladies (the states again). Un-

130

fortunately, the "side splitting and heart touching drama 'Loyal' that never grows old and always catches young and old, soldier and civilian" was rained out.[65] A special dramatic event intended to regale the veterans at Ash Grove was the burlesque capture of Jefferson Davis, but enough Confederate veterans were present that the play aroused some animosity. "There are tender spots which can very easily be touched on such occasions," cautioned the correspondent, "and precautions ought to be taken not to create any hard feeling among fellow citizens."[66]

Fortunately, grudges and hard feelings were not the rule. "The Yank and Johnny buried the hatchet, a thousand fathoms deep never to be resurrected," bragged the *Taney County Republican* on August 18, 1898. "All are now Americans and comrades in Taney County." Another editor noted that "the key attitude seems to be one of harmony and forgetting the past,"[67] and the orators on these occasions did their best to fulfill this mission by stressing the greatness of a united America. At Ava the speaker "delivered an address which touched responsive chords in the hearts of his listeners and more closely united the veterans of the North and the South which show that sectionalism was wiped out and we all stand on one common level and are ready at all times to take up arms in defense of the old flag."[68] At West Plains the Reverend O. W. Crow "discussed our relations to the other nations of the world in which he drew comparisons as to military strength, intelligence, national honor, etc. He was instructive and used excellent language in which to express his many thoughts."[69]

The old soldiers' encampments died out with the veterans they honored, but old settlers' reunions have continued into the present time. Both of these occurrences were large outdoor events. Only camp meetings could keep an Ozark farm family away from home for a longer period of time.

OCCASIONAL PICNICS

Throughout the Ozarks hundreds of picnics, large and small, were held each year for a variety of reasons or for no good reason at all. Sunday-school conventions, family reunions, lodge gatherings, and political rallies often took the form of picnics, permitting their followers to enjoy the Ozark outdoors, dinner on the grounds, and perhaps a program.

Republican rallies, Democratic rallies, Populist rallies, free-silver rallies, gold rallies—such meetings gave political speakers ample opportunity to have their moment, or hour, on the platform. Many smaller picnics welcomed such speakers, for they constituted an easily available, free, and always plentiful supply of talent to the program committee. The Peace Valley picnic was a favored platform for the politicians, for they could partake of the famous hospitality of that community and address twelve to fifteen hundred constituents besides. Thirteen candidates availed themselves of the opportunity in 1896.[70] In 1898 their speeches "were enjoyed as the boys were inclined to be more or less humorous and were in good spirits."[71] At Isabella, in Ozark County, "the festive candidate was numerously represented. He spared no opportunity to ply his favorite occupation, otherwise known as electioneering."[72]

In 1898 candidates of all parties were invited to a picnic at Pomona (Howell County), sponsored by the Women's Christian Temperance Union and were extended an invitation to address the people there assembled on the "issues of the day," which the sponsors identified as "temperance, morality, or anything that tends to promote social purity and elevate society." A candidate who accepted the opportunity would have found no gambling devices on the ground but could have amused himself while awaiting his turn to speak by listening to recitations and singing

The members of a Congregational church at a church picnic in Lawrence County, 1901. Courtesy State Historical Society of Missouri.

and sipping the "good pure ice water" that was free to all. The moral uplift of the occasion was given a further boost by the playing of the West Plains band, "composed of some of the best boys in the city."[73]

One year earlier the Salvation Army of West Plains published a heated denial that, as had been publicized, it had any part of a picnic that was to be held at Dripping Springs. Such amusements as baseball, horse and foot races, and dancing were to be on the program, and a spokesman disassociated the army from the affair: "The Salvation Army has too much respect for itself and the cause it represents to even think of participating in such things. . . . It is needless to say that no part of the Army will take part in the proceedings at Dripping Springs."[74]

The picnic at an old sheep ranch west of Galena featured a minister discussing the topic "Hard Questions and Easy Answers." Those attending learned the answers to the questions "Did man come from a monkey?" "Where did Cain get his wife?" "Did the whale swallow Jonah?" and "When will the world come to an end?"[75] The grand basket dinner at the Tumbleson homestead on Rippee Creek, in Douglas County, drew a crowd of about two hundred people prepared to hear the featured orator of the day, a black man, the "Hon. Thomas Foolman," who was also scheduled to make a balloon ascension. The "Hon. Foolman" turned out to be a local resident in blackface. The enormous success of his stunt was attested to by a correspondent:

> After dinner the colored orator Mr. Foolman appeared on a donkey with his balloon for the occasion. The crowd of about two hundred people went wild over him. The children flew in all directions when he fell over among them. After speaking for a short time he told the people he would go up in his balloon and take his donkey along. He accordingly lit his gass when unfortunately the balloon took fire. Mr. Foolman who it

134

seems was prepared for every emergency made a leap into the air and doubling himself amidship settled down upon the balloon and extinguished the conflagration in a manner that would have made a fire brigade ashamed of itself much to the amusement of the audience.[76]

Balloonists, real ones, thrilled audiences throughout the Ozarks with their spectacular ascensions and daring parachute leaps, and such events were features of many picnics, as well as fairs and circuses. They were evidently common enough to be taken for granted, and beyond mere notice of their occurrence newspapers made little comment about the balloonists' feats. Some notice, however, was give to an attempted ascension at Ash Grove which rivaled the "Hon. Thomas Foolman's" in spectacle and confusion: "They fired the thing up but couldn't make it work and after the man who was inside taking care of the machine became overcome with the gas and smoke so that they had to cut into the balloon and take him out it was abandoned by the show people and then it took fire and went up in smoke."[77] The people who had gathered on this occasion to see Madame Hiler of South Africa "leave the balloon in mid-air and descend in a parachute"[78] were sorely disappointed. Madame was a monkey, and the prospect of seeing a primate plummet through space would have been an appealing change from the usual leap by a mere human.

Ozarkers turned out for picnics, and they had a good time. Eating, talking, listening to speeches, riding the swings, or rooting at a baseball game, they would have agreed with the Clarkston correspondent in 1894 when he summed up his account of a picnic in the words, "In fact it was all good and the people all left happy."[79]

Gainesville will have one of "ye old time" barbecues on Saturday, September 15. Reps pops and dems will all have their say. All the county candidates will be present. Everybody come.

—OZARK COUNTY NEWS, SEPTEMBER 6, 1894

Other Entertainments

An Ozarker seeking theatrical entertainment could gratify his desires by attending the many special events that were almost certain to be held in his community throughout the year. These included political "speakings," baseball games, band concerts, community Christmas Trees, court weeks, and hangings.

POLITICAL SPEAKINGS

In 1896 from the small Barry County community of Golden came a one-line cry: "Golden for God and the Republican Party!"[1] penned by a reporter caught up in the fever of approaching election day. This equating of politics with religion gives some insight into the seriousness with which the Ozarkers viewed their politics and some indication of the joy with which they viewed an approaching election. The Ozarks were overwhelmingly Republican. Only Texas County consistently—and against the Republican tide running in the region—voted for the Democratic candidate in the six presidential elections between 1885 and 1910.

A political speaking meant the gathering of a crowd and the fellowship, conversation, and excitement that such an occasion inspired. Many Ozark political rallies drew audiences of 1,500 to 2,000 and often lasted for two or three days. Barbecue was often supplied free to the crowd, and individual families supplemented the entree with bread, cake, pies, and other picnic fare.

Troublemakers were not welcome at the Campbell's Station "barbecue and division of time"* in 1894. The announcement of the event warned:

> Eggs, especially rotten ones, will be boycotted and the first fellow who pulls a gun or shoots off anything but his mouth will be forever read out of the grand association of candidates. The first fellow found with a bottle of red liquor in his boot leg will be turned over to the deputy marshalls and the federal fee machine.[2]

At Ava in 1892 music by the cornet band, a series of baseball games for the championship of southwest Missouri, and "a score of other amusements to entertain the people" supplemented the speakings by candidates at a two-day picnic and barbecue.[3] A Republican rally at Mansfield in 1888 was hailed as "the largest gathering of this kind ever held in this section of the state." Estimates of attendance ran up to 3,000, and the highlight of the affair was the presence of the governor of the state, who reviewed a parade composed of three marching bands and 488 mounted men.[4]

Circuit court was often adjourned during a campaign period, so that political speakers could use the courtroom. A Democratic candidate gave an address of "liberal length" to a fairly well filled courtroom at Forsyth in 1899. The Republican newspaper reported that "he told at least one truth, viz., that he was no orator,"[5] but failed to report on the content of his speech.

---

*"Division of time" meant letting anyone speak who wanted to. The available time was divided up among the speakers.

In the fall of 1902, Mayor James A. Reed, of Kansas City, spoke at Licking to a gathering that was large for harvesttime. The editor of the *Houston Herald* was enthralled by the mayor:

> The speaker is an orator, one who can amuse, interest, and entertain. At times he held his audience spell bound. Again, tears would be brought to the eyes of his hearers as he eloquently portrayed the condition of the common people under Republican high tarriff and trusts. All united in stating that this was one of the best speeches ever made in Texas county, and the audience responded with applause and enthusiasm.[6]

In 1904, United States Congressman Champ Clark delivered an address in Ash Grove to a crowd that had to meet at a street crossing when it was discovered that the opera house, in which he was scheduled to speak, was already occupied by a traveling show.[7] Ten years before, in 1894, Vice-President Adlai E. Stevenson, passing through Ash Grove on a train, had been greeted at the depot by a crowd of several hundred citizens. Stevenson made a few remarks and then introduced the congressman of the seventh district, who was accompanying him. The congressman tried to speak, but a passing freight train drowned out his words. By the time the freight had rumbled away, the train bearing the distinguished party was pulling out for Springfield, while the frustrated congressman was trying to complete his interrupted remarks.[8] In an editorial in the Ash Grove paper the editor took to task some members of the depot crowd for being "either so thoughtless or so lacking in all the elements of good breeding as to 'hurray for McKinley.'"[9]

The people of the Ozarks were very much concerned about the issues involved in the presidential election of 1896 between William Jennings Bryan and William McKinley. The free-silver issue, establishing a two-metal monetary standard and permitting the "free and unlimited coinage of silver at the ratio of 16 to 1," was of great interest to Ozarkers, at least in part because the leading spokesman of the movement for nearly twenty-five years, Congressman Richard P. ("Silver Dick") Bland, was from the Southwest Missouri town Lebanon, in Laclede County. Bland was one of the foremost candidates for the Democratic nomination for president in 1896 and led all other candidates on the first three ballots. Only after Bryan stampeded the convention with his emotional "Cross of Gold" speech did Bland withdraw. There was enough popular (and Populist) support for the free-silver issue in the Ozarks to cause a number of voters to desert their beloved Republican party and vote for Bryan—the only time in a twenty-five-year period that the Ozarks supported a Democratic presidential candidate with their popular vote.

Free silver was debated throughout the Ozarks, and although most correspondents tried to be unbiased in their reports of these debates, most failed. A *Cassville Republican* reporter noted objectively that in a debate between a Populist and a Republican on the free-silver question "the points made by both sides were loudly applauded." His politics showed through, however, as he concluded, "It was evident before the debate closed that the adherents of the free silver fallacy realized that they had less to stand on than they thought."[10] Even more enthusiastic was the correspondent from Stony Point who mused: "How any man after four years starvation can vote for four years degradation we are not able to comprehend. Vote for McKinley and save our country from destruction. Hurrah for McKinley!"[11]

Excitement and spectacle characterized the large rallies, but the smaller meetings, such as those in rural schoolhouses, were met with no less enthusiasm and genuine interest: "Meetings have been held in most of the school houses and tired farmers forget the toils of the day and listen till after midnight to free silver speakers. Only last week one debate between a gold and silver champion continued until two A.M. and then some of the crowd rode five miles before they reached their homes."[12]

One of the most colorful and most fiercely partisan newspapers of the Ozarks in the 1890s was the *Cassville Republican.* Its editor delighted in excoriating Democrats, Populists, and a coalition party tagged the "Demopops" (or "Popocratsy"). During the presidential campaign of 1896 he declared on September 3: "We are taught that Christ was wise, honest, and a protectionist. Hence, McKinley and Hobart would head his ticket and it would be cast undefiled by blot or blemish." He and his correspondents delighted in ridiculing the opposition with mock prayers that they asserted were uttered at the Demopop meetings:

> Oh Lord, we the mongrels of a happy union formed by bringing the two greatest of thy mortal creation together which was done through the divine wisdom and foreknowledge of thy counsel from which we have sprung, the most noble of all thy handiwork, we beseech thee, Almighty Being, to prosper the most noble act of all thy life in bringing us any light at this critical hour, to save this our country from the clutch and grasp of the hateful Republican party. Oh Lord, we beseech thee to let thy wrath to rest on that hateful party, and to let mildew, rust and blight to attend them in life. But oh Lord, do let Heaven's richest blessings rest on thy chosen elect before the foundation of the world. Oh Lord, do show to the world that thy mongrels are of thy chosen people by giving us our immortal Bryan to be our next president, and Adam Herd to be our next Eastern judge, and all between them. These blessings in the name of the immortal Tom Watson and Prince of Peace Sewell, forever and ever, Amen.[13]

When Theodore Roosevelt spoke in Springfield, Greene County, in 1912, the *Springfield Leader* reported that "incoming trains were taxed to their capacity to handle the people coming to Springfield from every direction for a radius of more than one hundred miles." Courtesy Museum of the Ozarks.

Upon McKinley's victory the paper erupted with huge bold, headlines proclaiming the election of "The Apostle of Patriotism, Protection, and Prosperity," and although some of the extreme fervor seemed to leave the paper after the excitement of the election of 1896, it retained a fiercely Republican complexion for some time to come.

The period preceding an important election was a time in which almost all other modes of entertainment were subordinated to politics. Picnics, literaries, and even religious meetings took on a political flavor. And when the election was over, entertainment capital was made of victory: "A grand ratification, jollification meeting will be held at Ava Saturday night, November 17, consisting of bonfires, torchlight procession, short appropriate speeches, jubilee songs, and a general good time and don't anybody forget it."[14]

BASEBALL

In early autumn of 1892 the Mountain Grove baseball club notified the Ava club that it would be "unable" to play a scheduled game. After announcing this fact, the editor of the Ava newspaper wrote what started as a good-natured jibe but developed into a bitter harangue against the Mountain Grove club, startling in its vehemence in light of the triviality of the provocation. Of the cancellation he said:

> That is a wise step. They have outlived their usefulness, they have exhausted their "hoodoo," have met and vanquished, cheated and bluffed several amateur clubs, had the egotism to cross bats with ball players and were used as mops, chewed up and spit out and in the vernacular of the day discovered that they were "not in it." Don't submit your citizens to any further humiliation, gag yourselves, take your ball bats and herd geese, and the first man that pokes his slimy

142

head through the green scum of obscurity and yells
"baseball," make an angel of him on the spot.[15]

The editor's comments demonstrate how seriously Ozarkers took their baseball games. They were more than casual
athletic contests: on the conduct and performance of a
team rode the pride and prestige of an entire community.
The importance placed on the game assured the fans a
real spectacle each time the teams took the field. As a
matter of fact, being a spectator at a turn-of-the-century
baseball game was not necessarily a passive occupation. In
1902 in the eighth inning of a game between Lebanon and
Marshfield, in Webster County, the umpire called a Lebanon player out. The crowd poured angrily onto the field
to protest: "For half an hour they indulged in the most
violent, abusive language to the umpire and visiting team
and so great was their rage that for the space of nearly
half an hour there was imminent danger of a riot."[16] The
Marshfield newspaper credited the Lebanon team with the
"utmost sincerity," but suggested, "If one has the desire to
see a tough crowd that does not know the meaning of the
word courtesy, he has only to go to Lebanon to find it."[17]

Tension between the two teams' supporters was long-
standing. In a game three years earlier an official decision
had favored Lebanon. The Marshfield writer had been
pained when, with the score tied six to six in the eighth
inning, "the umpire seeing that his club was likely to be
defeated, by his rank and unjust decisions, deliberately
gave the game to Lebanon."[18]

A Gainesville umpire protected himself from over-
zealous fans and angry players by wearing two "mountain
howitzers" (horse pistols) buckled under his arms. In spite
of this, the West Plains team protested a decision early in
the game and, according to a correspondent, "kept on the
kicking act through the game, proving they could out kick
an Ozark co. mule."[19]

143

The *West Plains Daily Gazette* (Howell County) called it a "hinckey-dinckey" baseball game when it was played on the Fourth of July, 1900. The Cardinals, in individualistic "uniforms," won over the Blues, in equally colorful attire, 23 to 22. Courtesy *West Plains Gazette.*

Occasionally two clubs met in a spirit of peace, harmony, and sportsmanship. When this occurred, it was reported as a rarity: "There was not much quarrelling or wrangling as is usual in most games, and the Seymour boys say they were treated with respect and speak words of high commendation. They had a pleasant time and were royally intertained."[20]

The most colorful team that played in the Ozarks was a travelling professional club, the Nebraska Indians. Well promoted before they played the local clubs, the players were characterized as retaining the customs of their "fierce ancestors": "They refuse to go into a game without donning their war paint and chanting their favorite war songs. They carry their own wigwams. . . . They are said to coach in their native tongue and the old fairgrounds will be enlivened by their war whoops on the day of the game."[21] Such publicity drew large crowds eager for a look at the warriors in action, but in at least one town the people were sorely disappointed. The Ava editor reported: "There was only two that looked anything like Indians and two or three with course curley hair looked more like 'Dagoes' (Italians) or Negroes. A majority of them were white men."[22] Although it was a "great game and well worth going to see" (the Ava club won), the failure to produce nine full-blood Indians as promised lessened the theatrical impact of the afternoon.

Baseball playing on Sunday caused distress in some communities. A Houston correspondent admonished ballplayers:

> The Lord made this world in six days and on the seventh day he rested and while you are playing ball you are not observing the Sabbath as the Bible says you should. What a beautiful old world this would be if there were no such doings. Now young man, next

Sunday go to church and Sunday School* and when you return home see if you don't feel better and not so tired as the Sunday you played ball.[23]

In 1904 Truman Powell predicted to his Stone County readers that "the baseball craze is dying a natural death. It has done some damage, broken up a Sunday School or two, crippled some young men, but it generally furnishes its own remedy and we hope it will soon be over."[24]

In 1905 the Missouri legislature failed to pass a bill that would have prohibited Sunday baseball.[25] The Sunday games continued, and the crowds continued to watch and participate both vocally and physically in this dramatic sport. After the first five years of the new century, however, some of the gusto and frontier spirit seemed to fade away. The lessened importance of the game stilled such voices as that of a Salem, Arkansas, editor who, after his team was defeated by a Gainesville team in 1895, had written: "If we were forced to choose between Hades and Ozark county, Missouri as our future destination, we would study long and loud before making the choice for we figure the odds are rather in favor of Hades."[26]

BAND CONCERTS

Many an Ozark town boasted a band that played at public gatherings and marched in parades. The typical band gave a weekly concert in the summer months, when the appeal of music under the stars drew large crowds from town and the surrounding countryside:

> People from several miles round came in to hear good music and well they do for after a drive in the open air and listening to a good concert they return home in better spirits than when they came.[27]

*In the Ozarks most Sunday schools were held at two or three o'clock in the afternoon.

146

The concerts have been untold pleasure to the country people as well as the people of Ash Grove. Time without number have we seen wagon loads of young people from our neighboring school districts come to Ash Grove to hear the band concerts. The music was of a high class and was greatly appreciated by all who attended.[28]

Many towns built bandstands for open-air performances and gave concerts in the local opera house. Most bands were outfitted in military-type uniforms, the money for which had been raised by subscription, a benefit concert or pie supper or play. In 1877 the Cassville Cornet Band presented a benefit performance of *Milly the Quadroon, or, Out of Bondage*.[29] The war drama *Spy of Gettysburg* was quite a success at Marshfield, clearing more than forty dollars for uniforms. The newspaper reported, "The boys are so enthusiastic over their appreciated efforts, that they have decided to produce another play just as soon as it can be gotten up."[30] Special uniforms were donned for special occasions. A photograph of the Ash Grove band taken in 1898, during the Spanish-American War, shows the members wearing star-spangled trousers, striped swallowtail coats, and "Uncle Sam" hats, in preparation for the Fourth of July parade. The Gainesville aggregation dressed as a hayseed band at a literary in 1895. The costumes were varied, some players wearing "stars and stripes, other light searsucker coats and jackets with dark overalls, other with ragged coats, sloutch hats, etc. Large yellow string ties and mammoth watch chains added to the odity of the band's appearance."[31] The "Boer Band," which performed at the Harvest Home picnic in Howell County, delighted the crowds at that affair, sponsored by the German Catholics of the White Church neighborhood. The band played "America Mine Vaterland" and other patriotic selections, but it was their costuming that was the big attraction:

147

The Ash Grove Crescent Band, dressed patriotically in red, white, and blue stars and stripes for a Fourth of July performance in 1898, during the Spanish-American War. Courtesy Jeanette Musgrave and John Hulston.

They were dressed in uniforms, Boer style, with broad brimmed hats of dove color with a turkey feather ornament in each, all supplied with horns made of paper, not the cheap toot toot horns but made exactly like the regular band instruments from cornet to basso, and queerest of all they would play a tune. The band was led by two ladies in blue calico dresses cut short at the bottom in true South African style, showing low cut slippers and white stockings.[32]

It was a mark of pride for a town that its band was invited to neighboring communities to play at picnics, speakings, parades, and other gatherings. For transportation the band acquired a brightly painted and gaily decorated bandwagon, often pulled by four horses, so that the members could travel in style, rolling impressively into the midst of a gathering, uniformed and grand, playing a stirring tune. An amusing story was told about such a trip made by the Mansfield Cyclone Band. A farmer's organization known as the "Wheel" was holding a large meeting at a nearby town, and the band members decided that it would be neighborly to drive over and help out with the program: "As they approached the meeting place the band started playing with great gusto. As it happened, one of the prominent speakers secured for the program was right in the midst of his address at the time. The Wheelers were angry at this interruption and set out to overturn the wagon and band men."[33] Only by fast talking were the bandsmen able to convince the disgruntled Wheelers that they had intended no offense and had come only to contribute to the entertainment.

The one-line notice "Band Concert tonight" continued to appear in the weekly papers for many years after the turn of the century, and the town bands continued to furnish programs of music and entertainment for appreciative audiences.

149

Members of the Alton Brass Band (Oregon County) aboard their bandwagon. From Lewis A. W. Simpson, *Oregon County's Three Flags* (1971), p. 90. Courtesy State Historical Society of Missouri.

COMMUNITY CHRISTMAS TREES

Residents of small Ozark towns seldom had Christmas trees in their homes. Instead they joined in a community project, commonly called the "Christmas Tree," which featured a literary or musical program and the climactic event, the revealing of a large Christmas tree, handsomely decorated and laden with presents. Christmas Tree programs were held in churches, schoolhouses, courthouses, or any meeting place big enough to hold the large number of people who attended this popular event. May Kennedy McCord called the affair "the greatest and most thrilling event of the year."[34] Gifts for everyone, the children and adults, were on or beneath the tree, and the entire community took part in the distribution of gifts:

> As each beautiful doll was taken up from the tree each little girl waited with eager expectation to hear her name called, or when a beautiful album or picture was handed down the young ladies would wait with blushing cheek to know if they had been remembered by their sweetheart, and when the drums or guns were held up each small boy looked with eager eye to see if his name was on the tag. And the eyes of those whose heads were frosted over by the frosts of many winters twinkled with delight when their names were called.[35]

The tree was decorated with strings of tinsel, popcorn, and cranberries and lighted with candles. Presents hung from its branches and were piled on the floor beneath, evoking the admiration of all who beheld it. In 1893 the editor of an Ozark County paper squeezed his way through "a great throng of humanity," into the Christmas Tree at Gainesville and later wrote of the large cedar tree, "Its dazzling beauty and splendor as it met the gaze and admiration of thousands of eyes was something never to be forgotten."[36] As the time drew near for the gifts to be distributed, Santa Claus made his appearance and presided over

151

that phase of the entertainment, holding each present, calling out the recipient's name, and making witty remarks for the amusement and delight of the audience. Mrs. McCord described Christmas Trees held in the courthouse of her hometown, Galena. In addition to the regular presents, gag gifts were hung from the tree limbs, giving Santa Claus material for extemporizing "sight gags" and "roasts" at the expense of those for whom the gifts were intended. Among the gag gifts mentioned by Mrs. McCord were a diaper for a newly married couple, a stick of wood for the president of the local "never-sweat" club (so that he could be assured of holiday warmth without the necessity of exertion), and a generous cut of ham for the town's sole Seventh Day Adventist.

Local merchants subscribed to a fund to assure that every child would receive a present, and collections were taken up for special gifts. A Mount Vernon minister expressed his thanks for the $21.25 suit of clothes that was placed on the tree for him.[37] A newspaper editor thanked an anonymous donor for "a jumping jack of great beauty placed on the Christmas tree for him Monday night."[38]

Although the Christmas Tree received generally favorable acceptance throughout the Ozarks, apparently there were those who disapproved of the custom. Of these individuals a Greene County writer allowed that "it is their privilege to look at such things in their own way" but denounced them rather severely as "Puritanical, intolerant, and Pharisaical."[39] They may have disapproved of the tree as a pagan symbol. The desire to provide a more elaborate stage setting caused some communities to hold a Christmas celebration much as described above but to distribute presents not from a Christmas tree but from some other decorative device, such as a Jacob's ladder, a snow house, an arch, a windmill, an old log church, a Christmas wheel, or a fairy garden. The last was described as consisting of "boxes of candy, two large pillars mounted with floral urns, and four handsomely decorated trees."[40] Another celebration in the same town had as a setting "an antique fireplace and chimney with small trees on either side."[41] The New Church at Exeter struck a patriotic theme with a setting representing the battleship *Maine,* complete with sailor boys.[42]

The program that preceded the distribution of presents usually included "songs, declamations, and a dialogue."[43] It was a short program, for those in attendance were eager to get on to the more exciting business ahead. At a church Christmas Tree the children's choir often presented a simple cantata or play. In 1896 a performance of *Santa Claus's Reception* entertained the audience of the Union Church in Cassville,[44] and a large cast and fully costumed production of *The Trial of Santa Claus* regaled those in attendance at the Ash Grove Christian Church in 1891.[45] The drama, *The Christmas Pastime, or, The Crying Family,* given by the children of All Saints' Sunday School in West Plains in 1903, was well received: "The father, mother, children and all the months doing their parts well and when December came bringing Christmas and old Santa Clause, great was the delight of the children present."[46]

Much to the distress of the good people of the community, the Christmas Tree sometimes attracted elements who were intent on creating their own brand of Christmas cheer:

> We understand there was a very lively time at the Christmas Tree at Henry school house. A few fights, several eyes in mourning, and lots of tanglefoot.[47]

> We had very good order until it broke and then there was a right smart of shooting took place.[48]

> One man present said that he was going to kiss every lady present but was promptly knocked out by one of the ladies.[49]

153

> An old quarrel was the trouble which was renewed when they met at the Christmas Tree, and Levi shot at Elmer but missed him, no damage done.[50]

> The confusion consisted of swearing, hollowing, dancing, whistling, drunkenness, etc. right in the church house.[51]

Such conduct was not universal, and it was possible for many a reporter to compliment those in attendance for their demeanor, as did the Everton correspondent who noted that, in spite of the overcrowded house, "not a thing occurred to mar the pleasures of the evening,"[52] and the Gainesville editor who observed "the largest and most cultured and best behaved audience we have seen in a large time."[53] An Ava patron was pleased to report in 1897 that, "for the first time in many years at a Christmas Tree in Ava there was no drunkenness or rowdyism."[54] A Christian County correspondent wrote a glowing epilogue to the Christmas Tree in these words: "After all the presents had been distributed, the people started for their homes feeling life is not made up altogether of trials and troubles, and with a hope that they might be permitted to meet together on many more similar occasions."[55]

COURT WEEK

Twice a year, once in the spring and again in the late fall, the small county-seat towns of the Ozarks came alive with the excitement of court week: "Ava puts on a lively appearance this week. Crowds of people in attendance at circuit court. Merchants are doing a lively business and the lawyers are busy with their cases in court."[56] The convening of the circuit court was an occasion for the gathering of people from all over the county not only to conduct business with the court but to greet friends and neighbors and witness the drama of courtroom proceedings. "It was not

154

that they had any particular interest in it," said Emmett Yoeman, "but they didn't have any type of entertainment that we have today and so they had to devise something." "I would say that the Circuit Court contributed more to the entertainment of the people of the county than probably any other single event," said Marvin Tong.* "A term of court was a time when everybody came to town, and they'd stay as long as court was in session."

Hotels and boarding houses were soon crowded to capacity, and visitors who were unable to find lodging with acquaintances in town camped outdoors. At Gainesville they camped on the banks of the creek that flows in a half circle and borders the town on two sides. The broad James River bottoms accommodated court-week campers at Galena.

While the county seat was bulging with visitors, other towns found their population drastically depleted. In 1891 "nearly half the population" of Walnut Grove emigrated to Springfield to hear an interesting case,[57] while in Ozark County in 1898 a correspondent reported that "Thornfield was blank last week. Everybody gone to court."[58]

Juries were often selected from the men who happened to be on the streets at the time the sheriff was choosing a panel. Two men went to Ozark, Christian County, in 1902 on business and were summoned for jury duty and told to report at once. The sheriff was sent after them when they drove home instead.[59] Some families avoided going to town during the two months before court week for fear of being summoned for jury duty. Others, however, sought out jury duty and made themselves regularly available. Grover Denny boasted that "from the time I

---

*Marvin Tong was an Ozark historian and scholar and an editor of the *Ozark County News*. This comment and the anecdote below are from an interview with Mr. Tong in 1979 for the multi-image show *Sassafras*, produced by Southwest Missouri State University.

became twenty-one years old, I don't suppose I missed a year of some court a-bein' on the jury, till they thought I got too old to know right from wrong." L. O. Wallis told about the case on which Old Joe Pollack was a juror in the early 1900s. Before the trial began, one lawyer was heard to remark to his opponent that he had one juror on his side, that he had just loaned Old Joe Pollack $2.50. The opposing lawyer "popped his head down on his desk, said, 'I just loaned him five!'" Wallis concluded, "So I guess that was kind of a common practice in those days."

The real-life drama of a courtroom trial held a fascination for Ozarkers, whether or not they had a special concern with the case. With almost equal interest they observed trials for burglary, peace disturbance, horse theft, divorce suits, and murder. The man who beat his wife so violently that he wore out not only a buggy whip but a heavy limb from a peach tree as well, drew only a twenty-dollar fine on his guilty plea. If the women present from the surrounding community could have had their way, it was noted, "he would now be carrying a well-striped back himself."[60] Humorous incidents enlivened some trials, as at Greenfield, when the judge ordered the sheriff to call as a witness a man named L. G. Galligher: "The sheriff, being a witty kind of fellow, raised the window and yelled at the top of his voice, 'Let her go Galligher,' and at the third call the window came down on the back of his neck and reforced [sic] the broken dignity of the court."[61]

Although they enjoyed a good speech of any kind, Ozarkers especially favored the flamboyant, emotional style of argument used by most trial lawyers of the day.[62] Observers at the murder trial of Hosea Bilyeu had a special dramatic treat when the defense attorney, G. Purd Hayes, impersonated one of the murdered men: "Hayes dressed himself in the clothing, trousers, shirt and coat said to have been worn by Jimmy Bilyeu on the day of the massacre on which he gave up his life. Lawyer Hayes presented a striking spectacle, dressed in the bloody bullet ridden clothing as he appeared before the jury in the garb of death. The spectators . . . held their breath in silence."[63]

In an interview Marvin Tong described how an accommodating judge provided entertainment for an Ozarks audience during a winter term of court:

> They were having this murder trial and it'd come down along about 2 o'clock in the afternoon and it was time for the final argument before the jury with a couple of colorful lawyers. One of the natives in the back of the room raised his hand and asked if he could speak to the judge, which, you know, was sort of unusual, and the judge said yes, so this big fellow came up to him and told him, he says, "Now judge," he says, "it's going to be terribly terribly cold tonight." He says, "It's turned bitter cold out there." He says, "You don't reckon you could put off them final arguments until after supper time and then we could all come back to the court house where it's warm and spend the biggest part of the night here in the court house listening to the final arguments." So the judge said yes, he would. So he recessed the court. So everybody went back to their wagons and got supper, and then about dark why they came back to the court, and it reconvened and they were just hanging off the rafters in there. Everybody that could get in. And, of course, the lawyers just went all out 'cause they had a tremendous audience, you know. But this was a great source of entertainment and it was up until the late 1940's.

The appeal of court week as a special countywide holiday was due not only to such dramatic trial scenes but also to the special activity that went on outside the courtroom. When court was not in session, the spectators visited with seldom-seen friends and relatives, played games, and attended horse races and baseball games.[64] Refreshments were hawked to the crowd, and the local merchants ran special sales and remained open late at night to accommodate the rush of business. Local dramatic troupes worked up plays, and medicine shows were sure to arrive at the county seat.

HANGINGS

Executions by hanging were still public at the turn of the century, and to these gruesome events flocked men, women, and children from miles distant. Three thousand individuals flooded the little town of Ava in 1897 to witness the execution of a convicted murderer, Edward Perry. Many began arriving the night before, some having walked ten or twelve miles, and those who did not stay in town camped nearby. Of the huge throng only two hundred people were able to squeeze inside the enclosure around the scaffold to see the actual execution. Others paid for vantage points at windows and on the roof of the nearby Adams Hotel. To avoid disappointing the large crowd, Perry was taken to the bandstand, where a minister gave a talk on his behalf. Then the sheriff stepped forward and said, "Ladies and Gentlemen, I now introduce to you Edward W. Perry!" Perry's address, which was delivered "in a steady clear voice that could be distinctly heard all around the court yard," was short and apparently straightforward, though his statement "Good people of Douglas county, this perhaps will be the last time you will ever see me alive, but I hope to meet you hereafter" was a bit enigmatic. Concluding his remarks with "May God bless you all is my prayer," Perry retired to the scaffold and was hanged.[65] Afterward his body was laid out at one end of the courthouse corridor, and the people filed by to view it before returning to their homes and taking up their everyday lives once again.

Emmett Yoeman remembered the occasion from his

The Ava, Douglas County, courthouse, where Edward Perry was tried and convicted and where his body was displayed after he was hanged in 1897. From J. W. Curry, *A Reminiscent History of Douglas County, Missouri, 1857-1957* (1957). Reproduced by permission.

childhood and the intense excitement in Ava. He recalled being lifted up by a stranger so that he could see over the heads of the crowd. The man told him, "This is something you'll want to remember." Reflecting, Yoeman believed that "people came for the thrill of the thing and the excitement attached to it. A morbid entertainment . . . ."

The execution in 1887 of Edward Clumb in Barry County was witnessed by about seven thousand people, some of whom had traveled fifty miles or more to be present. Nearly a third of them were women and children. The hanging was apparently all business, with no public speaking by the condemned man for the entertainment of the public.[66] Likewise, the execution of three Bald Knobbers at Ozark on May 10, 1889, drew large crowds, but the few who were admitted inside the twelve-foot fence surrounding the gallows witnessed a bungled job of hanging (the knot on one of the nooses slipped, allowing the condemned man to fall through the trap all the way to the ground), but no entertainment provided by the prisoners.[67]

At his Texas County hanging in 1906, Jodia Hamilton entertained his audience of three thousand with a song and gay repartee: "The condemned man's appearance on the gallows was very dramatic. On Hamilton's appearance the crowd asked, 'Is that you Jodia?' to which he replied, 'I told you I would try to be here.'" Hamilton then made a rather long speech, which he interrupted to sing several verses of a song that included the chorus,

> Only a prayer, only a tear
> Oh! If sister and mother were here.
> Only a song to comfort and cheer
> Only a word from the Book so dear.

Afterward he continued his address. Another statement followed the prayer by the minister, and then the trap was sprung on the loquacious Jodia Hamilton.[68]

The hanging of Edward Clumb in Cassville, Barry County, on April 15, 1887. From *Keepsake Stories of the Ozarks* (Cassville, Mo.: Litho Printers, 1973). Reproduced by permission.

*There are some hoodlums in the community who put themselves
in all gatherings and never fail to show off even in church.*

—CASSVILLE REPUBLICAN, OCTOBER 27, 1898

Audience Participation

T HE Ozark orator who could not arouse his audience to
shouts of approval and encouragement might well consider his speech unsuccessful. The preacher who failed to
compel his congregation to cries of "Amen!" might declare
that "the spirit was not among us." The pie supper that was
not marked by good-natured wrangling and joshing might
be thought a social failure.

The Ozark audience was not a passive one. It assumed
an active and vocal participation in the entertainments
that it witnessed, and its enjoyment of an affair was thereby
greatly enhanced. Performers expected this manifestation
of interest and welcomed and encouraged its expression.

Political and religious meetings seemed to call forth
particular enthusiasm from their audiences. At a Republican rally held in Seligman in 1896, a correspondent reported that "peace and the very best of order prevailed"
and proceeded to describe the behavior of the crowd:

> The bursts of applause were frequent and long,
> and the echo of one prolonged cheer would scarcely
> cease to reverberate through the room before the pent-

162

up enthusiasm of some other old gray-haired voter would give way in a general footstamping and hand-clapping which everyone present seemed to want to indulge, and you could easily tell from the way that their feet went down on the floor and the glitter in their eyes that they felt just what they were manifesting. When Will E. Wallen, the Republican nominee for Recorder, appeared on the floor, their zealous fervor seemed to be completely enkindled and everybody went wild in a rapturous and ecstatic "Hurray for Wallen!"[1]

L. O. Wallis said that sometimes the shouting would "pretty near break up the service," and Uncle Joe Cranfield recalled the "shoutin' and prayin' and singin' and convictions" of the camp meetings. Mrs. Dunlap told about a little boy, performing at a literary, who concluded a "piece" he was reciting with a question: "And don't you think I've done well for one so small?" A patron shouted from the audience, "You sure have, boy!" Mrs. Dunlap also related that when the performance of *Mrs. Jarley's Wax Works* was completed a patron called out, "Let's hear that mermaid again!" and the performers obliged with an encore of "Down at the Bottom of the Sea."

There was, however, one group that participated in entertainments in a manner that pleased neither the performers nor the audience. This group, although decidedly in the minority, managed to impose itself upon a sufficiently large number of Ozark gatherings as to cause respectable citizens concern ranging from mere annoyance to physical fear. Emmett Yoeman carried throughout his life a scar across his nose, cheek, and chin as a vivid reminder of a knife wound he received during a disturbance at a box supper that he, as the teacher, was sponsoring. His predecessor had been run off two years earlier by hoodlums who had shot up the school building. Yoeman

A group of young men posed as Ozark hell-raisers in Branson,
Taney County, about 1906. Courtesy Douglas Mahnkey.

added in a joking tone, "But our school wasn't a rough school there. It was all right. But Lone Star, now that was a rough spot."

Claude Hibbard said cheerfully of Douglas County, "They used to kill a few down in this country," and when Fred Steele, of Hurley, was asked whether he had attended literary meetings as a boy, he replied: "Occasionally we would have literary meetings, but they didn't last long. At that time the boys in the community didn't get along too well. There would be fights and disturbance and usually it didn't last long. They would get in trouble and they would close it."

It is interesting to note that most of the older settlers interviewed denied with some heat that there had been any such disturbances in their communities. "Now over in the next district," ran a common reply, "they had trouble." It is likely that these old-timers were sincere in their protestations that their communities had been free of troublemakers. They knew that most of the citizens were peaceful and polite in a gathering. The number of disturbances may have been large, but the number of disturbers was small.

Rural correspondents for the weekly newspapers were great boosters of their communities and seldom missed an opportunity to stress what a fine group of people they were privileged to live among and how progressive their area was. Many of them found it newsworthy to report on occasion after occasion that "there was not the slightest disturbance during the evening" or that "perfect order prevailed." More often than not no mention of conduct either good or bad was made, probably indicating a peaceful event. When they did report disturbances, particularly at a religious meeting or a literary, it was usually with indignation and a moral lecture, in the belief that publicizing the wrongdoings would somehow shame the miscreants or arouse the community to action.

From the same correspondents, however, comes the most clearcut evidence that Ozark entertainments were often beset by troubles. The scribes wrote with a delightful candor about all the happenings in their communities, both good and bad. Names were used with abandon and a disregard for libel laws that would cause a modern newspaper editor to blanch:

> The hop at Mr. Richards resulted badly. Mr. Richards caught Marian Pippens with his arm around his wife's neck and went for him with an old broken file. Pippens run out of the room and started for home on a run. He had Big Sugar Creek to cross and when he came to it he thought Richards was still after him. He was afoot and hit the water like a frog all spraddled out. The water run up to his armpits and he went three miles in that shape, wet and wetter on a cool, frosty night.[2]

Many aspects of the frontier way of life were still evident in the Ozarks of the 1880s and 1890s. Men and boys carried pistols, and "blind tigers" (illicit stills) did a booming business. Tempers flared quickly at real and imagined offenses and were settled promptly, usually without recourse to law. "Spring fights opened with a four cornered jolly at Livingston's yesterday,"[3] wrote a Cassville correspondent matter-of-factly, as if pleased that the annual sporting season had at last arrived. A young man in Ash Grove, in town with a visiting theatrical troupe, was so unwise as to "insult" two women of the town: "He was promptly chastized for the same shortly afterward in front of the post office book store by Joe Sisk. He received several thumps that was a reminder to him never to do the like in Ash Grove again."[4] In 1887 a writer from Vera Cruz blamed much of Douglas County's troubles on "too much gun, cards, horse racing, whiskey drinking, etc. along with frequent violations of law caused principally by the non-enforcement of the same against law breakers."[5] The editor of the *Houston Herald* called for more rigid enforcement of the laws: "When drunken men blockade the streets, disturb all passers by with their loud and profane language, ride their horses on the sidewalks and destroy or damage the property of lone women, . . . it is time they should be handled and handled without gloves."[6]

Citizens were not loath to take the law into their own hands when they felt that the authorized agencies moved too slowly. In 1895 seventy-five citizens of Barry County masked themselves and "proceeded to chastize the three sons of Ed Wells" by whipping them. The Wells boys' offense had been their "proneness . . . to concoct and circulate scandal," and they had recently insulted a respectable lady." They were given fifteen days to leave the county.[7]

Contributing further to the rough-and-tumble frontier image were undisciplined toughs who roamed many Ozark communities. Harold Bell Wright based one of his characters on the type: Wash Gibbs, an antagonist in *The Shepherd of the Hills.* The real-life counterpart of Wash Gibbs has been described by many a country newspaper correspondent. The following account is typical:

> The order on the occasion of a singing school was very good except on the part of four or five rough characters headed by Bart White who came loaded with bust head whiskey and forced their way over the door keeper who was only a boy and entered the room without paying the admission fee. After accomplishing this wonderful feat they seemed not to be satisfied and so on the following Sunday evening to further distinguish themselves they went to Scott school house where a prayer meeting was in progress and the boss of this brave little band jumped on the boy who was door

keeper at the concert to eat him up but unfortunately for him he soon discovered that he had bit off a little more than he could chew.[8]

It was such a climate that the attempts of Ozark communities to meet in peaceful assembly were often frustrated.

Most disruptions were of three kinds centering on whiskey and weapons; those undertaken by men and boys who simply took delight in disrupting gatherings; and those resulting from lack of training in the social graces or experience in social situations. While these classifications are by no means discrete, they avoid lumping thoughtless young men who talked to their girls through the church windows during services with the drunken hoodlums who broke up meetings by throwing rocks and shooting off guns and dynamite.

WHISKEY AND WEAPONS

A Texas County resident had just returned from a box supper at Grogan when he wrote in 1902: "I don't see why those who have to drink can't stay at home and not come to the church house to drink and yell. Try to do better next time boys and wait till you get home to drink your whiskey."[9] Mrs. Earnest Hair believed that the reason so many men and boys came to such meetings drunk was that there was no "proper place" for them to go. Local option had closed many of the taverns in the Ozarks. There were 4,970 "dramshops" in Missouri in 1906, over half of which were in Saint Louis. Only 58 taverns were operating in the twelve-county Ozarks area described in this book. By 1910 local-option elections had reduced the number of such establishments in Missouri by more than 800, with sinful Saint Louis still accounting for well over half of those remaining. The number of legal drinking establishments in the Ozarks had shrunk to 28. Bootleg whiskey was easily available, however, and those who drank it went where the crowd was—school, church, or dance. The result was often like the melee at the Bralley School House, in Douglas County, in 1901:

> Whiskey and blood flowed freely for several hours to the entire disgust of the respectable citizens of that part of the county. . . . Last Saturday night a crowd of young bloods came to the meeting with plenty of fighting whiskey both outside and in. It was not long until the fight and the whiskey began to work and the boys went at it fist and skull.[10]

Trouble was averted at a New Site literary in 1890 when a young man, "having imbibed too much whiskey and being rather noisy," was taken outside by his friends and tied to a fence with a halter rein until after the people had left for home.[11] That community was not soon free from disturbances, for almost nineteen years later a very thoughtful and concerned letter to the *Commonwealth* began, "We have talked the matter over and came to the conclusion that an article on the misconduct caused by drunkenness at our pie supper at New Site last Saturday night ought to be published." The writer describes some of the conduct: "Horses turned loose, saddles hid, school books carried away and scattered along the road, and boisterous and unbecoming language."[12]

By this time, perhaps, New Site felt that it had had more than its share of difficulties. In 1906 a disturbed correspondent had cried emotionally, "Shame! O shame! Then let us pray for the heathen nearer at home!" He was referring to the "contemptable, cowardly, thieving wretch and lower than the most loathsome reptile that crawls on the earth," who would "sneak out in the dark and rob a stranger's buggy of a whip, laprobe, and not satisfied with that deface the vehicles in other ways, and cut and destroy the rubber tires, destroy or steal hitch reins and turn horses loose with the vehicles yet hitched to them."[13]

Religious services were not immune to drunken disorders. A Sunday-school convention in Taney County was disturbed by "a lot of young men? who made themselves hideous by getting drunk, using obscene language, discharging pistols and disregarding the law in general."[14] Too much whiskey caused the Christmas Eve festivities at the Clinkingbeard schoolhouse, in Douglas County, to become "almost too lively to be entertaining to the respectable portion of the audience,"[15] and the Reverend H. Pinkston was forced to abandon his sermon at midpoint "on account of a mob of drunken hoodlums who were creating a racket outside near the church. A continual fire of pistols and shot guns near a church on Sunday in this age of our Christian era seems strange."[16]

Although the law forbade the carrying of concealed weapons under penalty of a fifty-dollar fine,[17] it was common practice to carry a pistol or a knife, and the combination of drunken anger and a ready weapon resulted in many incidents of violence. More drunken disturbances occurred at dances than at any other Ozark entertainment, and shootings and stabbings were common occurrences: "At a dance not far from this place a few nights ago we were informed by a party who was present that there was no less than twenty revolvers present on the persons of the young men in attendance."[18] One issue of the *Houston Herald,* that of December 29, 1898, carried stories of shooting affrays at three different dances. Two Taney County boys took on a "little too much of the 'oh-be-joyful'" and took snap shots at each other, but the only person injured was an onlooker, who took a ball through the vest pocket. "All is well that ends well," the correspondent mused. "Both parties we understand have settled their differences and are now warm friends."[19] No report regarding the status of friendship—or health—of the wounded bystander was noted.

Knives were, if anything, even more common than guns as concealed weapons. In Taney County a dance was briefly interrupted when a young man drew his knife and stabbed another in the back and on the cheek. The latter grabbed a chair, felled his assailant, and the dance continued. The correspondent reported, "With the exception of this little difficulty a pleasant time was spent."[20]

Dances were not the only settings for violence with guns and knives, however. In 1903 the literary society meeting at the Jameson schoolhouse, in Webster County, witnessed an assault in which a patron acting as sergeant-at-arms was shot in the neck, "the ball barely missing the main juglar vain." One can imagine the reaction in that crowded schoolroom: "For a time pandemonium reigned for supreme. At the time the house was filled to its full capacity and it is stated on good authority that those inside went through the windows like a wild flock of sheep."[21]

On October 9, 1902, a Hartshorn correspondent to the *Houston Herald,* in reporting a court case regarding the carrying of concealed weapons, fell to musing about the kind of person that was typically involved in such a charge. His analysis is most revealing:

> A young man when he gets old enough to have a nice suit of clothes and a jack knife in his pocket must have a revolver and go to a preaching or to a dance and if his word is disputed the least bit out comes his revolver with an oath and a threat to shoot. Of course his friends run to the rescue at once and that ends the row until another opportunity comes. Now, there is hardly ever anyone hurt or much harm done, still it spoils the harmony of the preaching or the dance. It gives the stranger the impression of a rough set and causes many of the better class of citi-

zens to keep at home from gatherings and never to give the young people an entertainment. Hence, carrying concealed weapons and not knowing the common rules of good manners mars our country, destroys every effort at higher social gatherings, and the good people stay at home and get rusty and the youths are left to their own devices to sink lower and lower on the social scale.

DELIGHT IN DISRUPTION

A letter to the editor of 1894 suggests the formation of a brass band at Gainesville. The letter contains this passage:

> In small places where music is not cultivated in any marked or public manner there are many opportunities offered to youth to spend their leasure time either in trifling ammucement or dissipation.
>
> The evenings are devoted to lounging about saloons and billiards halls or standing upon street corners while the habits that loafing inculcates are far from improving to the many young men who inhabits our smaller towns.[22]

The reporter might have added one more less-than-wholesome recreation of certain young people, even (or especially) in communities where public gatherings were frequent. That recreation was the deliberate and intentional disrupting of an entertainment, usually a religious meeting or a school affair. The evildoers anticipated community entertainments with just as much enthusiasm as did those who sponsored them, but for quite different reasons. Their whole delight was to mock and disrupt, and in their purpose they often succeeded to a degree that was maddening to the rest of the community. "If you can't act respectably, for God's sake, stay at home until you can,"[23] suggested an exasperated correspondent, and no doubt an annoyed community sighed, "Amen!"

A political speaker at Pleasant Hill, near Everton, was booed and "otherwise disturbed" by a group of hoodlums and was forced to stop before he had talked ten minutes. When he started to walk away, they threw rocks at him.[24] At Seligman a literary met "with closed doors on account of an experience that other teachers have had in the school with open doors."[25] At a church service in Stone County a woman was prostrated with fear when a group of boys rocked the church house; she had to be taken home in a buggy.[26] At Forest Grove "someone threw an empty bottle into the tabernacle meeting Saturday night causing quite a disturbance."[27]

Religious services were the special target of these rowdies, and while their methods varied, most were crude. One of the most common and unimaginative was to create such a racket outside the church that the preacher was compelled to stop speaking. Guns, black powder, rocks, and noisemakers were all employed in this endeavor. At Raymondville, after continued tapping on the window disturbed the services, the minister asked someone to go outside and stop the racket. Several members of the congregation did so, and a free-for-all resulted. Fortunately, it was reported, no one was seriously hurt.[28]

Mormons and members of the Holiness (Pentecostal) sect were especially persecuted. Two Mormon preachers, holding a meeting on a street in Licking, were assaulted by "unknown persons" throwing eggs,[29] and another Mormon minister was struck in the face with a rock while preaching at the Cross Hollows schoolhouse, in Barry County.[30] Earlier in the same year an evangelist and his helpers, holding a series of meetings at Buzzard Roost, were driven out with rocks, and the "young toughs . . . would not allow them to come back."[31]

A protracted tabernacle meeting held by the Holiness sect in Greene County was broken up by a band of ruffians: "While the meeting was in progress these parties cowardly stole up to the tents untying the ropes and they commenced to throw missles and stones at it. No one was injured and after the meeting was broken up these toughs undertook to intimidate these people by firing off pistols."[32] The Reverend F. J. Light, a retired Holiness minister who lived and preached in Howell and Wright counties, had "eggs and rocks and frogs and I don't know what all" thrown at him while he was preaching.

"Squig Hensley" agreed to be interviewed only after being promised that his real name would not be used. He poured forth a series of youthful exploits he claimed to have engaged in, some of which went considerably beyond the bounds of decency. Completely toothless, wallowing a wad of tobacco in his mouth, and wearing nothing but a pair of filthy overalls, he leaned forward in a rickety chair in his front yard in Ozark County, his eyes gleaming, as he recalled the entertainments of his youth and the great fun he had breaking them up:

> I went up there one day on a Sunday. I'd had some of that old mountain dew. I'd take a little whiff of it. I threw a saddle on my horse, and I thought, "Well, I'll have some fun today." I went up to the church and I shot that dad-burned pistol until it was red hot. I shot it empty, and I took out. . . .
>
> We'd go to meetin', a bunch of us boys down there, and we'd get outside the house and make a racket, anything we could, you know. One time we made us a dumb-bow.* Made it of old cans. Boy, you could hear it from here to Kansas. We'd go up there to meetin' and pull it slow, you know. And oh! that preacher'd jist look and stop and says, "Well, the Lord may be here from the racket they're makin' outdoors." . . .
>
> One night while church was a-goin' on, the old preacher he got down a-prayin' with a bunch of young

*A noisemaker, or "bull-roarer," often a resin-covered string pulled through a tin can.

174

girls, men, and women. And there was a nice young girl there, about eighteen year old, and that old preacher he ran his hand up that girl's old Santa Claus, and she just slapped him plum back on his toes, I hollered and I howled. That old preacher got up, and he said, "I was just a-doin' that to try your faith. Oh, glory hallelujah!" . . .

Well, they got up the Holiness Church. Well, when they got that up, it just suited me. I'd go up to the mourners' bench at night, you know, and I'd get down and have a big time. Them old girls'd get down up there, you know, and I'd have bad feelin's. Them girls'd get down on their knees around me, and I'd forget and let my hand slide down thataway and they'd put it back.

The experience Squig Hensley gained during the time he spent in religious places he later utilized when he conducted church services in the county jail for fellow inmates and admiring townspeople who gathered on the sidewalk outside.

A fine of one dollar and costs, amounting in total to over ten dollars, was the punishment meted out by a justice in 1888 on a guilty plea to the charge of disturbing religious worship.[33] That was the usual fine in the Ozarks for this offense, though more flagrant cases drew stronger fines. Thirty days in the county jail was the fate of each of three young men from the vicinity of Bakersfield who disrupted the Sunday peace of Ozark County,[34] and a judge at Ash Grove had his own formula for assessing a fine on the young man before him who "persisted in putting his feet over the seat in front of him and kissing his girl who was sitting beside him in church. Evidence given proved that he kissed the girl three times, for which the court saw fit to charge him five dollars each."[35]

Throwing rotten eggs at an unpopular performer was a time-honored tradition of wide acceptance and certainly was not peculiar to the Missouri Ozarks. Egg throwing seems to have been particularly common in the little town of Ash Grove, however. In 1889 the local editor was moved to muse that "this Ash Grove egg throwing is as mysterious as the London White Chapel Fiend" and deplored the rain of eggs that descended upon the sleight-of-hand performer and his audience as they were gathered in the street in front of Chandler's Store.[36] Three years earlier the same editor had been roused to wrathful denunciation of the parties who tossed the "volley of ancient eggs" at the medicine-show entertainers as they performed on their wagon. "The egg business has went to an entire extreme in Ash Grove," he thundered, "and this extravagant use of them should be suppressed."[37] The years took their toll, however, and by 1893, it is sad to note, no fiery editorial voice was raised at a recurrence of the offense. Only a short, factual note reported that "at the entertainment given Monday night by the Kickapoo Medicine Company some of the young boys indulged in throwing eggs in the crowd."[38]

LACK OF SOCIAL TRAINING OR EXPERIENCE

Most Ozark communities, even those which were spared drunken or chronic troublemakers, were embarrassed and harassed at their public gatherings by youth who behaved in what may be charitably described as an unrefined manner. The older members of the community deplored their conduct but believed that it was not motivated by maliciousness or a willful desire to disturb, but rather proceeded from ignorance of the basic social graces. Misconduct of this sort was embarrassing and distressing to those who cherished as a norm of behavior and breeding respect for religion, for parents, and for community:

> One of the best places to judge a well bred man
> or woman, boy or girl is in a church at the hour of

worship. A well bred infidel or sceptic will treat the minister with the respect and courtesy which is due even though the listener may not believe a word the preacher says. If those who laugh and grimace on such occasions knew how badly they looked and what other people thought or said of their conduct, they would for their own sake be more thoughtful.[39]

Boorish behavior most often took the form of boisterous and inconsiderate conduct and especially by the indiscriminate distribution of tobacco juice. Ash Grove College asked that those attending the literary exercises, "refrain from using tobacco in the building and spitting on the floor."[40] At Washburn,

> some vandals so far forgot their early training (if they ever had any) as to besmirch the walls and floor of our school building by spitting tobacco over them until it formed in pools on the floor during the literary exercises last Friday night. Vigorous application of a bogardus kicker would be of a lasting benefit to such people.[41]

A Douglas County correspondent condemned the practice of making "a spittoon of the floor of the house of worship." Of all the revolting sounds he had ever heard, he said, "the limpid 'splatter' of a mouthful of tobacco in the back part of the house of God is the worst."[42] Four young men who congregated in the northwest corner of the schoolhouse at Protem, in Taney County, disturbed an otherwise orderly and excellent literary program by "talking, laughing and dancing, spitting tobacco juice, and in other ways making themselves ridiculous in the extreme. . . . We are inclined to believe that ignorance more than a desire to disturb actuated these fellows and therefore hope that our grand jurors and prosecuting attorney will take no note of it."[43]

Correspondents often threatened to utilize the power

of their office and print the names of habitual offenders, but the threat was seldom carried out. Veiled or overt threats of arrest were frequent, though one editor recommended the more direct approach of administering a little "'peach tree tea,' which would do the kids more good in less time than all the courts in America."[44]

Young men often brought their young ladies to church but remained outside during the services, meeting the girls after dismissal and escorting them home. At one church at least such separation was no barrier to sociability. In 1897 a correspondent complained, "The thoughtless boys who slip around the outside of the windows and talk to the girls on the inside of the Baptist Church are disturbing those who assemble there for worship."[45] Even so, those who remained outside were usually considerably less of a nuisance than many who came inside. In 1906 the *Ash Grove Commonwealth* vividly described a situation common in the Ozarks:

> The habit of leaving the room by our young men and boys during Sunday School and church service is a lamentable fact. Young men will take their seats with their sweethearts and once or twice during the services they will singly and in pairs leave the room, walking heavily, loudly shutting the doors, creating a racket which is not only unbecoming to themselves but disrespectful to the minister and congregation. Besides it sets a bad example for the younger boys. Young men, what would you think of your best girl's getting up and going out from one to three times during services, and while passing from the room walking heavily and loudly, and while out hang around the door, talk loudly and boisterously, smoke, chew, and spit, and come lumbering back and seat themselves beside you? Some of you do this way. Too many of you. Shame![46]

In 1903 the *Aurora Advertiser* commented that "there are

times when it is necessary to leave church services and no one recognizes it quicker than the minister" but that gadding in and out a half-dozen times during an hour service was carrying necessity a bit far.[47]

A revealing composite portrait of this boorish breed is painted by a disgusted subscriber in a letter to the *Taney County Republican* in 1899:

> There is a type of young man in every community whose ambition it is to make other people tired and he usually succeeds. People often become so tired of him that they pray for his taking off in some manner and when his death is announced it is hard for them to keep from declaring a holiday. This young man is strongly in evidence at all public gatherings and often at private ones uninvited. He usually wears a sinister, smartaleky look, a speckled necktie, when the old man has any money, a pop gun, a pint bottle of extract of hell in his hip pocket. At public meetings he stands about in groups while the program is being rendered, dodging and craning about his inquisitive eyes, searching through window or door for objects of malicious comment, his eager ear strained to catch the remotest semblence of coarsness in jest or remark. When the program is well underway he comes klonking in, a dozen of him at a time, if his force is sufficient, and all sound of music or of speaker's voice is drowned until he is seated. He now spits a blob of tobacco juice on the floor, almost enough to float the benches, straightens back, thumbs in arm holes of vest, and stares out of countenance those about him. But he has not come in to remain. Just at the time when every sensible person is intent on catching every syllable that falls from the speaker's lips, our tiresome yet tireless young man, the same quantity of him that came in or possibly more, deliberately arises, and slowly klonks out again.[48]

The peace of community entertainments was disturbed

179

in many ways, some commonplace, some strange. "A little fistic encounter" during religious services at the Coffey schoolhouse in Ozark County was attributed to differences of opinion between the Christians and Baptists of that community, though the correspondent assured his readers that "the boys were 'scrapping' just for the fun of it."[49] Early McCollum used some "bad grammar" at the New Salem literary and apologized publicly for it at the next weekly meeting.[50] Emulus Grantham was arrested for disturbing the peace at a literary at Scott School House. His offense was "building a fire in the stove and making the room too hot to be comfortable, and by loud and unusual noise."[51] A Sunday-school convention was forced to compete with a swing for the attention of its young folks. The correspondent noted that it was the first time he had ever seen a swing at a convention and hoped that it would be the last. "Dear Christian friends," he exhorted his readers, "let us guard against the evil ones (swings) ever coming in the SS cause any more for the great day of His wrath is coming and who shall be able to stand?"[52] Perhaps the most entertaining diversion of all was reserved for a Sunday-school convention that met in Barry County. The Dog Hollow reporter relates the incident in splendid allegorical style:

> After the interesting exercises of the convention it was learned to the amazement and regret of all good people that a so called "soiled dove" claiming to have flown from Springfield had lit in the brush near the school house and by her cooing was attracting the attention of a certain uncertain class of young men. But no such conduct could last long in this community, for upon the proper complaint being filed before Squire Patton, the soiled dove was soon coyed out of the bush by Constable Amos and soon escorted to the pigeon hole in Cassville, there to mourn while waiting further developments.[53]

180

It is easy to accuse Ozarks audiences of being rude, crude, and boisterous, and there is no doubt that by more sophisticated standards than were common in that area, such terms would often not be incorrect. But a more tolerant and objective view would hold that the great majority of the audiences were well behaved, deeply interested, and very proud of their local performers. The norm of acceptable behavior called for active and vocal participation, and that the audiences gave, while any disturbances or trouble at entertainments was caused by individuals who, in relation to the total audience, were very few.

10
Afterthought

THE hundreds of thousands of tourists who flock to the Ozarks each year, creating bumper-to-bumper traffic jams in many of the popular resort areas, would find it difficult to comprehend the isolation that once characterized the region. Thanks to highways, airports, and electronic media, modern Ozarkers have access to the same cultural and entertainment events that the residents of other areas of the country enjoy. Young people entertain themselves in video arcades, at the movies, and before the television set. It is an easy trip to Springfield, Tulsa, Kansas City, or Saint Louis to attend a symphony concert, a ballet, an opera, or a play.

Yet while Ozarkers today do not find it necessary to gather at the schoolhouse, the church, or the picnic grounds for entertainment, much of the older culture is being preserved and is flourishing, encouraged by, of all things, the rise of tourism.

Many of the popular country-music shows are comprised of family units, and the instruments they play are guitars, fiddles, banjos, and dulcimers, just as they were played by music groups of long ago. A hundred years ago Ozark craftsmen produced handmade items essential

182

An early tourist in the Ozarks, R. W. Wilson (right) was a salesman from Kansas City who moved to Taney County to become a major land developer in the area. Courtesy Lynn Morrow, Center for Ozarks Studies, Southwest Missouri State University, Springfield.

Today old skills are being relearned, and young Ozarkers are preserving the crafts of their forefathers. Here a young blacksmith demonstrates for tourists how to forge-weld a chain. Photographed at Silver Dollar City by Terry Bloodworth.

for everyday living. Today these skills are being relearned, and a new generation of Ozarkers is practicing the crafts of its forefathers. Blacksmiths, woodworkers, quiltmakers, chair caners, broom makers, basket weavers, and candlemakers produce handmade items in the old-fashioned way to sell to visitors to the Ozarks.

For the past few years it has been possible to buy, from a local tourist enterprise, a log cabin, constructed of hand-hewn logs held together with cut nails and wooden pegs and completely furnished with hand-crafted furniture and accessories. The price for this rustic reminder of our past? Something over $100,000. Included in the price is a two-place log outhouse, and construction on your lot.

Ozarkers grind flour in water-wheel gristmills, lead tours of the caves where Bald Knobbers once gathered, and act in pageants and plays that re-create life in the early Ozarks.

While most of these ventures are obviously commercial, the financial rewards they provide have enabled many of the crafts and much of the culture of the older Ozarks to be preserved. The modern physical Ozarks is a far cry from the isolated, sparsely settled land of a century ago. But many of the characteristics of those earlier residents— independence, self-reliance, conservatism, pride in home and community—are preserved in the people today. And these latter-day Ozarkers continue the tradition of making entertainment for themselves and for others from the activities and artifacts of everyday life.

Appendix

Transcriptions of Selected Interviews

INCLUDED in the appendix are transcriptions of some of the interviews that I conducted with older residents of the Ozarks in 1960. The residents were encouraged to reminisce as they wished. The interviews were recorded on a portable tape recorder, and the transcriptions have been edited only to remove irrelevant or extraneous materials. Grammar and syntax, colorful and unique expressions, and colloquialisms are retained intact, though little effort has been made to indicate peculiar pronunciations, elisions, or contractions. The interviews are in alphabetical order by speaker. The place of residence given after each name is the speaker's residence at the time of the interview.

Chick Allen *(Stone County)*

A brush arbor was built in a grove as they would say, and they would clear out, and there would be a lot of little trees, and they would cut poles and put up just like the framing of a building. They would cut a great big forked sapling to hold that, and then they would take poles and crisscross that, and then they'd cut brush and lay on top of it. In a hard rain they would leak, but to keep the dew off or a little rain or the sun and so on, you see. You might say they were a big shade.

A big brush-arbor meeting would be a camp meeting. People would just go there and camp. Stay there day and night. For seating and all like that they would just saw off blocks in the old days. They would just have a bunch of blocks and just lay lumber down on it. Everybody would bring a quilt maybe to put on that seat.

There was a lot of wonderful old-time preaching in them days. As far as dictionaries and high-falutin' words back in them days, they didn't know that. There was a world of them preachers that couldn't even write their name, but they would maybe have somebody to read their text, you see, and otherwise they would just get up there and take off. Otherwise it was just them and the good Lord, that's all. Them educated preachers back in that day and time among the hill people here, they just didn't exist. On Sunday and the like of that they would have a big basket dinner, all the people who lived in the area with services at noon, and it would just all be spread out together there. Whatever was there was for everybody.

Well, maybe you lived over here five mile, and I'd see you maybe once a week. We'd meet on the road, and I'd throw my hand up at you, but we couldn't set and visit. Well, back in that day and time, visitin' and a good fellowship with your neighbor and your friend, that was it. We didn't have places to go. We couldn't get up and go like they do today. In that day and time ten miles was a long way, and twenty miles was a hard day's drive with a old lumber wagon, with the roads we had.

A way back at fox hunts it was more or less fiddle music. I was fifteen years old before I ever seen a guitar. It was fiddle or maybe a banjo. When I was a kid a-growin' up, a fiddle and a banjo and a pair of knitting needles and the jawbone of the mule like I play, and if we had a square dance a woman might happen to have an organ. There was a few organs in the country, that was all the music. This jawbone of the mule is the underjaw of an old dead mule, been dead for years and years, and you can beat a rhythm on that just the same as you can on a drum. I've played an old jawbone of a mule for over forty years all told. All I ever do play any more. I've got a big old pencil I beat it with, but hit don't make any difference. You can use an old rib bone. You can use the knitting needles to beat the time on a fiddle string. (Out on the creek you can get these skunk

190

burrs and let them season and use them. Knitting needles hurt your fiddle strings.—*Mrs. Allen.*) You need a steel knitting needle. You can use copper, but copper has a more dead sound, or you can use sticks, but they don't have the volume that a steel knitting needle has. Our band wasn't like any other in the world.

Walter Baker *(Springfield, Greene County)*

Down there at the old Theodosia [Ozark County] in the olden days, the General Baptists would have their camp meeting each summer. Well, they would bring their wagons. Now, that was before the day of automobiles. They would bring their covered wagons, and they would park them all around. They brought their food, and they would either sleep in the wagon, or they would have tents, and they would spend about a week there. They called them General Baptists' conventions. Now the General Baptists, as you know, is different from some of the rest of us Baptists in this respect—they do foot washing. They have annual camp meetings and general revival efforts, and the gist of this is that there was a feller down there, he was a distant cousin of mine, and he and his cousin were great practical jokers, or impractical jokers. So during this meeting one night they saw these ladies nursing their babies, most of them suckled them, of course, at the breast at that time, so they saw them putting them back into the wagons, here and there, and they said, "Wouldn't it be funny to swap those babies around?" Well they did it. Now they swapped the babies around, and it wasn't long till one of the babies cried, of course, and momma came out to that wagon. She looked in there, and that wasn't her baby, and she screamed, and the other mommas ran out to their wagons, and it wasn't their baby and they all started screaming, and the men was so incensed, so angry about it that they said, "If we find the fellows that did that, we'll lynch 'em." They were Baptists, now, but they were going to lynch the boys that did that. They had a hullabaloo time, broke up the meeting for the night. They finally got the babies all swapped back.

In about 1895 I started to school. At our school we had entertainments of sorts, Punch and Judy shows, traveling, and a little bit of magic once in a while, magic shows. My old grade school was Oak Dale. It has ceased to exist. Country school twelve miles west of Gainesville. I know a boy who lives over here. He and another boy made his way to the Pacific Coast and back by doing magic. That was the kind of shows it was. These little fellows coming through just doing one-night stands.

We had ciphering matches, spelling matches, and we had literary efforts. That was just grade school, and in high school we had a continuation of the same. When I was in Ava High School, we had a principal there named Dameron. He was a huge, angular fellow, reminded me of pictures of Abraham Lincoln. Mr. Dameron was a good teacher and a fine fellow. He had a huge, angular face with a mouth as large as you ever see, and one of the students, Friday afternoon at the literary efforts, Charlie Dean, says, "I have a poem." "All right read your poem," he said.

> His head is like a teakettle
> His nose is like the spout
> His mouth is like a fireplace
> With the ashes all raked out.

And the superintendent knew he was referring to the principal. No names were given, but the superintendent knew he was referring to the principal. "Charlie, Charlie! That'll do. Sit down." Made it worse.

We didn't have the parents Friday afternoon. Occasionally one or two. But when you had spelling matches or ciphering matches at night, then they would come out pretty well. Our schools would compete. They would go from one school to another and compete in spelling and sometimes in ciphering, but mostly in spelling. They would give catchy words. "C-o-n, con; s-t-a-n, stan; t-i, ti; n-o," and everybody would yell "no, no, no," and that confused the speller and he stopped. Jokes like that, you know. But ofttimes they did spell by syllables.

Near the end of the year—those were short, my first year was five months—but near the end of the year they would always have a literary program and invite all the parents. And we nearly always had a treat for our students. It would be either candies,

192

apples, or something like. But we always had something for our students at the close of the year, and they were always quite disappointed if we didn't. No playing cards were allowed; and, when I was teaching at Hammond, Missouri, I had two eighth-graders who wanted to use playing cards at noon and recess, and I said, "Well, I think that the patrons would object, so I guess we'd better not." It made them so angry they left the Hammond school and went to Gainesville.

Disturbances often happened down at church. Some of the boys would get ahold of some White Mule, moonshine liquor, and they would raise a disturbance in various ways by yelling and talking outside, once in a while throwing rocks against the church building. If they got caught, they paid a pretty heavy fine or maybe got locked up for a few days. People were very sincere about not having religious services disturbed but the boys generally were good.

There at Isabella—the old nickname was Goober—Isabella is on Highway 160 between Gainesville and Forsyth about half-way. I had two cousins, twins. They were small fellows, even now they wouldn't weigh more than a hundred and thirty pounds each. They were at a public gathering—it seems to me it was an entertainment there at Goober—and somebody insulted Cousin Herman's sweetheart, and Herman invited him outside, and he came out, and he was near twice Cousin Herman's size. And, of course, Cousin Herbert, his twin, came out to see that things went OK. They got into a double scrap, and Herbert's opponent got him down and was choking him and beating him, and he was about twice Herbert's size, and those twins was doing their best, and Herbert reached into his pocket. He had one of those old Barlow knives with a blade about two and a half inches long and it has been used around the farm. It was just like a saw, almost. It wasn't sharp at all. We called 'em chicken callers, they was so loose in the handles they would click. So Herbert pulled that out, and he started raking the other fellow around the cheek with it. With his left hand he was holding his opponent. His opponent commenced yelling, "Take him off! Take him off! He's killing me!" And the opponent was on top all the time. Well, to prove that my relatives weren't at fault, those other two boys were arrested and put in jail and had to pay big penalties.

Mrs. Queen Bell *(Mansfield, Wright County)*

I was raised over close to Hartville. We had spelling matches. We had singing matches, different singing classes. We had singing teachers that would go around over the country, teaching singing school maybe two weeks. We would meet at different churches or different schools. We had singing matches just for the fun of it, between two classes. It was all just for fun. There wasn't no prizes or nothing like that.

We would get the old farm wagon and fill the bed with hay. We'd go places that way. We'd go to our singing schools that way.

We would play all kinds of games. Sometimes it would be what they called parlor games, and sometimes it was a running game. They would sing songs, you know. Instead of having music like you dance by, they'd have a song, and they'd sing that song, and they'd play "Skip to My Lou, My Darling." That's what we'd call a play party, in a house, you know. It would look quite a bit like a square dance, but there wouldn't be no calling off or nothing like that. Just the song, you know.

Spelling bees would usually be two different schools, and they would meet at school houses. Each school would be against the other one. They were pretty well attended, mostly young folks.

We had debates, and we had literaries more often than other things.

Lots of time at night if there was a protracted meeting somewhere also, a bunch of us would get together in the old hay wagon and go seven or eight miles away. Mostly we had protracted meetings, but there was one place, called it Pleasant Valley, they always had a camp meeting every fall, but that was quite far from us, and I never did attend that. They had a whole bunch of little houses like we have for traveling people now, you know, and people would go and do their cooking and all right there. It was usually about two weeks. We were in Wright County. We had brush arbor meetings too.

We were four miles from Hartville then, the way the road went. Now it might be a little over a mile. We didn't go to town

194

very often. My dad never believed in going places unless you had business. On Saturday he usually went to town to buy what groceries we had to have. Of course, we raised most everything except coffee and sugar and things like that. On Saturday was his day for going to town. I never went when he went in that way unless there was a show in town. Sometimes there would be a show in town, and they'd have animals and performers. They would give riding in the ring, trapeze, almost anything. After I was grown, in my last teens or first twenties, I went to Hartville to learn to sew, to trim hats. While I was there, they was very often a medicine show come. Of course, we always went to medicine shows and had a big time. They would have performers that would sing and talk and advertise their medicine. I don't think there were any traveling plays or anything like that around here. Just a little old circus and medicine shows.

They always had court the first of March. That was the biggest court they would have, was in the spring. In the fall sometimes they would have a trial of different things through the year somehow. We wouldn't go to town for about two months before court week for fear of getting on the jury. My family never did attend court. The lawyers were from different towns. They come here. Of course, they didn't have the conveyance they have now so they would come and stay till their part of the court was over.

Mrs. E. T. Brown *(Gainesville, Ozark County)*

As I said, there were two sessions of court: November and May term of court. And that meant that people from all over the county came. They would come in covered wagons and camp on the banks of the creek, which is on two sides of the town—it makes a kind of a half circle. And these people that come in these covered wagons year after year, they would select the same camping spot. On Sunday afternoon they would begin gathering, and my crowd, my age, would get on our horses and go down by all the camps to see who all was here. . . .

They held all the criminal-court trials in the courthouse, and it would just be jammed to the walls with people, with spectators. We'd all get together and go up to watch them feed the

195

prisoners in jail. People usually began going home on Friday. The lawyers would come from Springfield. Judges from Ozark and Ava.

Uncle Joe Cranfield *(Kissee Mills, Taney County)*

I lived in Medleyville [about seventy years]. . . . We'd meet together when we was young folks, kids, we had a bunch at my Dad's nearly every Sunday. . . . The first school I ever went to the teacher got $20 a month, and he had a family and they lived on it. . . . Oh, once in a while we'd have a box supper, and I had my first fight at a box-supper night. Never did know what it was about. . . . Well, we went to this box supper that night. 'Twas in the fall of the year, still warm, and some neighbor boys was a-standing out in the yard in front of the schoolhouse, about like from here to my house. Well, you know how kids would do. This boy come up. The crowd was between me and him as he came up. We had been a-goin' to school, and there wasn't anything wrong that I knowed anything about. He come up, and we spoke to him. He never spoke to me. I noticed it at the time. He come on over to where I was standing, and he just shoved out his left hand at my breast, and he shoved me back, and I said, "What's the matter with you?" I says, "Cut it out!" and he struck at me. The next thing he got up about as fer off as them blocks out there. Well, that was all he wanted. Well, the rest of the boys out there, they began to put on like they wanted to take it up, and I said, "Hop on, I've got plenty here for all of you." But they didn't want none of it.

Well, we had one debate up there once about the nigger in the South. A debate about the darky how he was treated and the other side about how well he was treated. Well, one boy got up and he was just a-makin' this as he went. He said, "Them darkies didn't get nothin' to eat down there when they was slaves, only rolled oats, and they hadn't never been shelled—they just got trash and all together." Well, that pretty near broke it up, and he lost over that. They'd debate sometimes on the prettiest girl.

Sometimes they would speak some pieces or have literary doins before the pie supper, ever once in a while, not too much. A lot of times they would have a jar of pickles for the lovesick couple. A good cake for the prettiest girl. Sometimes they'd go wild about that. I've knowed that cake to bring as high as thirty and thirty-five dollars. Hard on the man that bought it but good for the school.

I'd give a hunnert dollars right now to go to a real old-fashioned camp meeting for thirty days! Every preacher around in the whole country would come. It didn't make any difference about his denominational standing. He was "Brother So and So." It didn't make any difference. They left off their denominations. They preached the Bible. They'd gather up there. I expect I've seen as many as forty or fifty preachers at one of them, and I've knowed of them a-lastin' for thirty days. Well now, before this would start, the men would go out and hunt a place out in the woods. It wouldn't be too hard to get to. And they'd cut poles and forks and build a scaffold up overhead there, and then they'd cover that with brush for shade. I've seen as much as an acre covered that way, and I've seen a crowd that was big enough to fill it full. And of all the times that ever you saw! Well, you never saw such times as they'd have—shoutin' and prayin' and singin' and convictions. It'd just make your hair stand up. We'll never have it no more. People have got out of that notion. Some old farmer off maybe twenty miles would load up his wagon with grub and horse feed and drive in there and he'd stay till his grub played out and he'd go back home and get another load. Maybe he'd go back and get a second and a third load.

They'd usually put their stock in the pasture. They didn't worry about nothing when they went to church. And the idea was with everything that went was to get right with God. They wouldn't have no disturbances, they wouldn't allow that. They didn't seem to want to. The toughs didn't get very fur down there. They just didn't get very fur. . . .

Now the Mission Association, they come to Dry Branch once. They was over four hundred wagons and buggies come. And that looked like—you would be up on some of them hills, they'd be down in the valley, where you could see the whole business of the campgrounds, and them camp fires just looked like stars. Well, they wasn't a-doin' much good there for a while one time, on this time where there was so many people. They had a preacher appointed fur the service that night, and he got up and told them, he

197

said, "Now we've got about thirty minutes yet before preachin,'"
and says, "We'll have a devotional service." That's testimony.
You tell what you been a-doin' and how you felt about it and
what you wanted and what the Lord had done for you.

There was an old lady got up from Arkansas. She started
in. She said, "When I first went to Arkansas, there was no Christians down there that I knowed of but myself." Says, "I finally
got acquainted with a neighbor woman, a Christian. She belonged
to my church." Says, "We'd get out and meet in the woods and
have prayers. That was about all we could do, and kiss each
other when we'd separate." She'd tell that now for testimony at
the White River Association. She says, "And it was like pulling
water over a pulley. We could feel that spirit come down."

That thing broke loose, and they got that testimony meetin'
started, and it lasted till after two o'clock, and they was thirteen
conversions came out of that testimony. The preacher couldn't
do nothing with 'em. He tried to stop them, but they wasn't
nothin' doin'. They was testifyin' all over that place. Thirteen
conversions come out of that. Well, there is a lot of people now
that don't believe in God or anything else, only get your dollar
if you got one, and if you ain't, why give you a kick and send
you on to get one.

Well, they'd take those converts to the river and baptize
them, and then everything broke loose when that happened.
They'd shout, them old women, and the old men, they'd go wild
too, a-hollerin' and the tears just a-pourin', not of grief, of joy.
Like the Lord said, "Joy unspeakable and full of Glory." What
can you find any better than that? They's nothin' any better
than to get into contact with God and stay there. . . .

Back in days gone by they wouldn't baptize if the water was
awful muddy. They thought that it should be clean. And if a
baptizin' come up, and say the water was muddy, if they couldn't
find a clear place, no filth, they would just say, "Well we'll have
to wait a few days." If you was goin' to wash your clothes, you
wouldn't go into a muddy stream to do it.

I've seen 'em break the ice to baptize. Well, now, a lot of
people would think they'd freeze to death, and they don't. I
never knowed a one to take any cold over it. Some times thirty
or forty would be baptized at one time, and of all the shoutin'

and screamin', singin', prayin' you ever heard, they'd have it. Didn't hurt 'em none.

We'd have some of the awfullest times there ever was. This old man Harrison Hall he put in a case in the kangaroo court about this neighbor that he'd stole Aunt Bess Garth's old sow. Well, they was just ahavin' it around and around. First thing Harrison Hall knowed they slipped in a witness that swore that Uncle Harrison Hall stole Aunt Bess Garth's old sow and took her off, they seen him agoin' with her. That just like to aruint 'em out at the kangaroo court.

Docia Davis *(Mansfield, Wright County)*

I was raised on the Gasconade River, seven miles north of here. They had camp there then. Pleasant Valley they called it. They had camps there then. Buildings, you know. It was built for that. The C.P. [Cumberland Presbyterian] church had it. It generally lasted a week, sometimes two weeks.

What little schooling I've got I walked a mile and three-fourths to get. I crossed the Gasconade three times. We had literaries, and we had spellings. We had big schools then all the way from fifty to sixty in one room, one teacher. The first year ever I went, I went three months, and after that they got it up to four months, and then the last few years it was six months. The whole entertainments on Friday afternoons was the spelling. It was mostly just for the kids. Part of this time they had literaries on Friday night, but I never went to them. My father was very strict.

On the Fourth of July that was a great thrill going to see the Calithumpians. They'd be dressed up and generally have a clown. We generally went to Hartville. It is only about five miles from my place to Hartville. They'd have a parade, you know, picnics, and once in a while they would have a balloon ascension.

Grover Denny *(Cabool, Texas County)*

We went places in bunches. We didn't do like they do now, just one in a car. I was raised three miles this side of Mountain Grove [Wright County]. I walked two mile and one half to school. We

199

had entertainments at the schoolhouse and pretty near all the schoolhouses around there, spelling bees and ciphering matches, box suppers and pie suppers. We had lots of music and recitations. All the people of the community worked together to put on the entertainments. We'd take an old wagon and put three or four spring seats in it, and we'd all go in a bunch. We'd sing and just raise hell along the road. Nothin' meant by it. We just had a good time.

We had no brush-arbor meetin' at all because we could always go to the schoolhouse or over at the church. About four families there built that church. It was a community church. We didn't have no General Baptist—we had a community church, but it finally drifted off to the Methodist church and then the Baptist church stole it. We had a regular preacher. We went and enjoyed ourselves. We had as good a Sunday school, better than they have now. Each class didn't have to have their age. We'd have four or five classes there in that one room, but we all got along. Our preachers never got up there and preached for money. They preached for the good of the people. I can go to one church just as good as I can another one, and I think I've been around enough to know right from wrong.

From the time I became twenty-one year old, I don't suppose I missed a year of some court a-bein' on the jury, till they thought I got too old to know right from wrong. I lived there, and Houston was a way over there. We had a long way to go. A lot of them went just to see what the other fellow was a-goin' to do. They knew there was going to be lots of talk. They was going to put on a show over there. They went to see the whole trial. They went to listen and see how that fellow was going to come out.

There was a community spirit. We had no cause to cross people. We all worked together. Of course, there was Democrats and Republicans but you have got just as much right to your belief as I have to mine.

I've seen good baptizings, must have seen twenty or thirty baptizings, and we didn't have these pools to baptize in. We went to the creek. The Brethern they call them now, we called them Dunkards, they had a church just north of us, and the old men wore the long whiskers and the old women wears little bonnets, and if they told you they'd pay you a dollar today they'd pay

200

it. The Dunkards come from the East, and what [nationality] they was I don't know.

A lot of people would come just to see a baptizing down at the river. Why, it would just be like going to a carnival. There would be people from all over the neighborhood there, and they enjoyed seein' you. No one was there to make fun at all; they was there for what good there was there.

We had debates, but I never did debate because to tell you the truth I never did learn language or history. . . . This country is about half-and-half Democrat and Republican. Course, you take these Dunkard peoples or these Baptist peoples, they're going to vote Republican or not vote at all.

Mrs. Earnest Hair *(Hurley, Stone County)*

In my memory Christmas was a pretty good-sized day but it didn't begin to compare with Children's Day. I remember one thing, it always meant a new white dress. The little girls always had their white dresses, and, of course, at that time the little boys wore knee britches, and they would try to manage to have a new pair of knee britches to wear for Children's Day. We had songs, and we used to, we would say speeches. We also observed May Day and May baskets.

When churches began observing Mother's Day in church, Children's Day just sort of got left out. Children's Day always meant more to me than Christmas even.

[At Christmas programs] they always had a sash or a broad band of some color draped across their shoulder and tied over on one side.

In Stone County, generally, there were no Christmas trees in the homes. We had a definite German influence in our community. My sister went to school at old Oak Hill School, which was straight up the railroad track four miles. In this little school was two families of Germans. One was named Livingston, and one was named Werner. The girls in these two families were good friends of my sister. She and her playmates from here stayed all night with them around Christmas time, with the Werner girl one night and with the Livingston girl the next night, and in each one of those homes there was a Christmas

201

tree. The people through here observed Old Christmas. That was probably in 1906, 1907.

We always hung up our stockings, and the few toys, whatever we had, we got them in our stocking without thinking anything about a Christmas tree.

For entertainment I doubt them going more than six or eight miles. Now, then, this thing of going places at night, I don't know whether you could say more than four or five miles. Mostly it would be horseback with the boy providing the extra horse for the girl. We went from here out to the outlying communities to social affairs before we had any place here for them. Nearly always there was a money-making element connected for some cause.

That was the day of the protracted meeting, and we didn't know the word "evangelist" or "revival." Those words were new words. We didn't learn them till much later. Camp meetings were held at Elm Spring, west of here. That is the fox hunter's paradise over there. In the period following the Civil War and previous to the 1900s, camp meetings were held over there. It came as almost a shock to think of these various meetings as being a form of entertainment, but it was. That was what you went for. Even funerals. People would talk about, "Oh, she just cried and she just cried and she took on something terrible," you know, and those people would just think about it for days afterward. It was a conversation piece.

When they were building the railroad, there were great gangs of Greek laborers and men of other nationalities who would camp all up and down the railroad. I was just a little girl, but I do have a definite recollection of that time because everything was so different. I remember the Duffield Troupers. They come on account of the railroad men, you know. That, I am sure, that was one of the first or was the first traveling troupes that come through Hurley. The Weaver Brothers came through here in about 1910. We had a big old empty room upstairs in the town they used. There wasn't any hall or theater in connection with it, just a big old empty room.

All we got out of court week was just the hearing. Someone [would] come home and tell about it. We would get it secondhand, but even so it was still a big thing. It was probably bigger than you realize, since neighbors were eager to hear about it. There was one thing here that was purely a local thing. We once had a bank here in Hurley, and when our bank was organized here, my dad was the first cashier of the bank. The first Monday in the month was bank directors' meeting. The directors were scattered around. One lived on James River at the arch. These fellows came in to the bank meeting on Monday. My dad would invariably bring two or three of the fellows home with him for dinner on bank-meeting day, and Momma'd get so provoked because it was washday, and she wouldn't have a very good meal, and she thought that was kind of belittling that she couldn't put her best foot forward on account of it was washday. Now I remember thinking, "Now I just don't see why Mom can't put her washing off on bank-meeting day and cook a good dinner," because it was a special occasion. My Dad never did talk a great deal, but if anything funny or anything amusing had come up in the conversation he would relate that at suppertime. The day that the bank directors came in and met was one of the big days of the month.

Someone said not long ago that the reason that these little country churches such as out in the country and in the little towns were bothered with the element that they once had to fight was that they are now going to the proper place, which is the tavern. Used to you could always count—during a protracted meeting or something to do at the church at night—you could always figger that there would be a bunch of roughnecks around the outside of the building. Always! I believe that that didn't cease until there was some other place for those people to go and be just as obnoxious as they had been around the church. They knew they weren't wanted at the church, but that would be the place that they would always gather. Almost always that would be the place that they would always gather. Almost always there was somebody causing trouble. But my, it has been a long, long time since anything like that has happened.

There was a little church close to the arch that they named Bloody Ridge. One fellow knifed another one, and he died because he lost so much blood. Those things just happened. You could just count on it, and in thinking about something, I remember hearing the older ones saying, "I wonder if so and so will be there with his bunch?" It wasn't the conduct on the inside you had to contend with. It was also the element on the outside. It didn't matter what church they belonged to because they seemed to just go from one place to the other in the little places around here. If there was something going on around here that they could stand outside and be a nuisance, they would, and then they would go into Union City or Black Jack or Walnut Grove or wherever and do the same thing.

Around here we had an old gentleman named Ped Miller, and he drove to his wagon a little mule and a big old horse, and the horse would always be a way out here and kinda carrying the mule along. Well, Ped did more than peddle pans and all the stuff that was in his wagon. He carried the news. We didn't have even here a daily newspaper, and the county weekly didn't have much local news, and about all that you knew was what you picked up from someone who peddled something, for instance, baskets. East of Hurley there was the Gotney family, and they were the basketmakers, and when Uncle Marion went from one farm to another as he traveled with his baskets, he would find out every single bit of news that he could. He would stay till he found out everything that they could tell him and then go on with his baskets to the next place and tell those people what he had learned at the other place and pick up any additional news. So these early tradesmen, peddlers and the like, served as news media. Selling their wares was their primary concern, but always they had a little bit more to offer, and people were always glad to see them coming. If the person had something to sell and just sold it like the Fuller Brush man does nowadays, I doubt if that man could have made a living. We always looked forward to anyone who traveled. There was one Syrian—no, his country was Lebanon—and his name was Jarvis, and he lived and died in Springfield. He started coming through this part of the country when he carried his pack on his back, and he worked up until he came through here driving a Cadillac. This doesn't sound reasonable, but it's true. When he first came through this part of the country, I was just a little girl, and my mother always bought a tablecloth from him. When he would come to your door, he remembered your name. He made it a point to remember all these people. He was one of the most interesting people that I have ever known. I looked forward to his visits, not only to seeing the pretty things that he had in his pack, but he would always tell something that had happened here or happened there. I think the part that the peddler played in the lives of women particularly would be hard to estimate. Women had so little to look forward to. My Momma never did even visit a great deal, and to me when we would see that old man coming, why I could have just hugged him because I was so glad to see him. His beads and stuff like that, Momma never would buy 'em or let us buy 'em, but he would always end up by giving us some little trinket, and she would always buy a tablecloth, and you know that it was real linen too. I think I still have one of the tablecloths she bought.

Mr. and Mrs. H. A. Hamilton *(Mountain Grove, Wright County)*

Mr. Hamilton: Came here from ol' Virginia in 1905. I lived north of Simmons town for seven years, and then I lived east of Simmons [Texas County] and then came back up to Nixa [Christian County].

There wasn't too much to do when I was growin' up except only dancin' and picnics. We had dances and speakin'. We only had three months of schoolin' when I was raised up, and I quit when I begun to get big enough to learn. I was twenty-one when I came to this country. We lived about two mile and a half from Simmons town. We went to town once a week, sometimes maybe twice. Done our tradin' there. We used to take off Saturday and go to town and prowl around a little while.

Mrs. Hamilton: We used to have programs all the time. I was in one program I think about. Went to the country schools all around here and went to Cabool. We said these things at the Friday-night literaries or at a pie supper. I was always in all those. When we came to a pie supper, the whole community would be there, and we looked forward to those pie and box suppers to have a good time. We called a pumpkin pie a "new-ground" pie because they always put out pumpkins on new ground.

205

Sometimes they would have a cakewalk at pie suppers. You would pay a dime to walk in the cakewalk, and then they blindfold someone, and, you know, drop a broom down in between the couples and whichever one they hit, why they got the cake. Sometimes the cake would bring, oh, a lot that way. We walked in the cakewalk not too awful long ago, Pa and me.

Well, you know, when we lived in the country, we didn't get to town in the evenings ever because we only had the horses and buggy or to walk, so we just didn't come in at night at all. We lived about eight miles out, and the roads was not very good. We had awful bad roads. We would walk across the hills to prayer meeting.

We had lots of brush-arbor meetings. People would come for miles and miles. And then we used to have the [Baptist?] Association, and people would come from miles and miles and camp there the whole three or four days. They'd be as many as a dozen preachers and singers come from other places. We really had a good time. I know we lived about three miles from the church house, and there was one old man that sang there, and we could hear him at our place of a night. He used to sing, "Farther along we'll know all about it, farther along we'll understand why." We could hear him just as plain. He'd preach awhile, and then he'd sing awhile like the old preachers used to.

Me and my brother went down to Bond's Lake, I don't know, it's way through yonder. We went through Tyrone and Lickin' and down in there. They had an old reunion there, and they had the awfullest crowd you ever saw. They had doll racks and fiddles. They had one man that put them doll racks out of business down there. He could just hit 'em. And fiddle! Man, them old long-armed men! Oh, talk about fiddlin'!

When I was baptized, I was baptized the twenty-second day of February. There was snow on the ground and ice on the crick, and they's thirty-three of us baptized that day, and I walked about as far as from here to the bridge down there just in my sock feet, and they cut the ice and moved the ice back to make room to baptize us, and it didn't make a one of us sick, and there were thirty-three of us.

We used to say "Preachin' on the grounds and dinner all day."

206

"Squig Hensley" *(Gainesville, Ozark County)*

There was some boys lived there at Pontiac, name of [?]. Boy, they was outlaws. Me and them boys did everthing we could think of. I went up there one day on a Sunday. I'd had some of that old mountain dew. I'd take a little whiff of it. I threw a saddle on my hoss, and I thought, "Well, I'll have some fun today." I went up to the church, and I shot that dad-burned pistol till it was red hot. I shot it empty, and I took out. First thing I knowed I was at their residence, and the law was after me. I didn't pay any attention to that. Them boys they would keep a tad tail. They'd watch for me and do anything they could about seeing that the law didn't get me.

One day they had a big decoration. I was about twenty year old. They had a big decoration there at Pontiac, and I just went right on up there. I got my wild hoss and all that. We went to lunch at twelve, had a table fixed there, a great long hunter's feast, and I was just a-helpin' myself. I had my hoss hid over there in the bushes. I met them neighbor boys Dell and Newt and said, "Boys, I want you to watch today and hep me with that law." I said, "There is so many people here they might get a-hold of me but," I says, "I'm going to eat some of this food." They said, "All right." I got standing up there at the table, and I was eatin' when directly one of them boys says, "Look out, Hensley!" I looked down about thirty feet, and here come them laws right down through the crowd a-pushin' and a-walkin'. I just dropped down and got off down the side of the hill there in the bushes. Watched him, and when he started back up there, I eased in behind him, and I was filling up my craw again. Well, they gave me two or three chases, but I got plenty to eat. I took a plate of chicken and dumplings and wandered off down the hill and helped myself. But directly one of the boys says, "Look out there is two of them coming around there, the deputy and the law both right down that side there." I just eased out, and when they started, they had to come around the whole crowd there. I knowed where my hoss was, and I knowed what it'd do, an' I just made a dive for that hoss. And when I made a dive for that hoss, that hoss just took off. He knowed something was up. I tossed the lines off the brush at him and I just jumped the saddle there and just tuck out down the road. Directly I heard something and looked and them two laws was comin' on that hoss and that hoss was just a leavin' em like a train or car leavin' a man. We had to go about two or three miles before we could hit the line, and I went over the line, and the old hoss stopped. I turned around in my saddle, and I looked back, and I said, "Well if you fellows will go home with me, I'll let you sit down and eat just like a man, not be like a dawg run off from the table." I stood there as long as they would talk to me, but they wouldn't come on over where I was at.

We'd go to meetin', a bunch of us boys down there, and we'd get outside the house and make a racket, anything we could, you know. One time we made us a dumb-bow, made it of old cans. Boy, you could hear it from here to Kansas. We'd go up there to meetin' and pull it slow, you know. And oh, that preacher'd just look and stop and says, "Well, the Lord may be here from the racket they're making outdoors."

They had the awfullest time in them old churches that ever I seen down there. They'd get up there and preach and sing and have a big dinner at noon. You talk about filling up, boy, I would.

One night while church was a-goin' on, the old preacher he got down a-prayin' with a bunch of young girls, men, and women. And there was a nice young girl there, about eighteen year old, and that old preacher he ran his hand up that girl's old Santa Claus, and she just slapped him plum back on his toes. I hollered and I howled. That old preacher got up, and he said, "I was just a-doin' that to try your faith. Oh, glory hallelujah!"

There was eleven families on the old river place. Well, they had dances ever night. There was a little mean fellow there name of Walker. They had a dance at his house one night, and they went out and got a gallon jug for a dollar there full of old mountain dew. So they all kept a-drinkin', and that little fellow four feet tall weighed a hundred and sixty pounds. So after a while he got teed up, and after a while another one up there got teed up, and him and this man name of Walker got into it, and he had an old razor and he jammed into him

208

twenty-two times straight up the handle. Well, that Walker fellow passed out, and they put him into the bed, and they went on with their dancin'. And after a while they come to find out he was about to float with blood all over and seen what a shape he was in. Well, that little dickens he made a getaway. He got up and hid in the ceiling till nearly daylight, and then he pulled out and they never did get him.

I remember there was an old fellow talking, he cussed all the time, and he says, "Yeah, whiskey was the whole goddamn cause of it. That was what was the matter." And it was, I guess.

Well, they got up the Holiness Church. Well, when they got that up, it just suited me. I'd go up to the mourners' bench at night, you know, and I'd get down and have a big time. Them old girls'd get down up there, you know, and I'd have bad feelin's. Them girls git down on their knees around me, and I'd forget and let my hand slide down thataway, and they'd put it back. I'd take and raise up and look at the preacher, you know. I had a bottle of fluid right here. I'd smell it—why you could smell it a mile seemed like. I'd smell that, and the water'd just pour outta my eyes.

Claude Hibbard *(Ava, Douglas County)*

Right on the Ozark County line about five miles north and west of Dora. Spent my early life between Blanche and Dora.

Girls were different then in one respect. They would walk a hundred miles at that time if you led her by the hand and walked with her. If you didn't have any other conveyance, they went with you, to a dance or a party or church. And if you had a horse, which we usually did, they wasn't above mounting behind you and holding on tight and going. All in all we had a good time. I never owned but one buggy in my life till I was married. The church was your outstanding social function because there you met all the neighbors that you hadn't seen for a week. And people went more universally to church then than they do now. They rolled up there in wagons and now and then a buggy. That was the very elite that had the buggy, you know, saddle horse, mares, and colts. After you got a little older, you either took a girl or took her home.

209

The church to me in my early life was very important. Not from the standpoint of my being so radically religious but from the standpoint of the influence it had on my life and the people who attended. There wasn't many social functions. Of course in every school we had a literature society of some sort. And that was no mean kind of training. We met on Friday evening usually. Readings, jokes, recitations, songs for those who could sing, and debates. And a little gossip paper. A lot of the English that children learned then was taught right through that. The questions they'd debate were questions that could never be resolved, most of them. Everything from slavery to present-day politics and almost whether the hen came before the egg or the egg before the hen. But the principal thing that came out of it was that youngsters and often grown people were a part of this debate. Youngsters developed the ability to stand on their feet before an audience and make a little contribution to the entertainment.

The dances of the early days were practically all square dances and later on round dances. The round dances came to our community by way of the German element. Most of the churches with very few exceptions banned dancing. Now my own parents didn't, but they didn't like to have us go because they felt that drinking and fussing came up at dances, and that was oftentimes true in some of the communities, especially after I was about grown. The dance was more of a private affair. Dancing was banned, but play parties were not, and at play parties the youngster did not have the violin music, but the youngsters just to the singing marched around and kept time to "Old Dan Tucker," and "Skip to My Lou," and "The Miller Boy," and such games as that. They even resorted to kissing games.

We used to have lots of revival meetings. I remember two ministers. One was a Christian minister, and the other one was Baptist. They got in an argument one night, and they had to hold them apart to keep them from fighting. They had the revival meetings, and the sermons were very fiery, of the brimstone nature often. Some were excellent preachers. They could give you something to think about, but most of them aroused the emotions. I think we would agree that a religion without emotion is no religion but that a religion that is all emotion is a very poor religion. And they were largely emotional, and people came up to the mourner's bench, as they called it then, and the church workers with the best of intentions wept with them and pleaded with them. Pointed out the eternal fire that they were destined to unless they repented their sins, and sometimes those meetings would get very happy and emotional meetings. I can remember some dear old ladies that would jump up and down and shout, and really they enjoyed it, and you could see that they were extremely happy. And I have seen men get happy, too. All in all at that time it was right good. We had revivals in school buildings, and then most communities built churches. Mostly it was the Methodists and the Baptists in this country.

The professional shows were almost nil. I remember when I was probably ten years old, a little circus, we thought it was a circus, it had a few animals, a high act — one fellow dived from a forty-five-foot pole — I thought about that for a long time afterwards. Had some snakes. It was a wonderful show, but taken now at nearly any modern park it wouldn't draw a trickle of youngsters. It was a wagon train. That was the day when the people believed that the only way you could settle a personal argument was by personal combat. They used to kill a few down in this country.

Tom Hogard *(Gainesville, Ozark County)*

We had literaries. Debating, you know, and dialogues and recitations and such as that. We had a kangaroo court here, and we would take up cases, you know. Sometimes the truth and sometimes a lie. We had a case here, tried a fellow for stealin' watermelons. He happened to be a lawyer and in the next day or two he had a case come up at Bakersfield defendin' a fellow for stealin' watermelons. For a fine maybe, they would have to set up. We used to have plenty of good liquor.

Oh, seventy years back they used to have camp meetings here. My grandfather was a Southern Methodist preacher, and I went with him when I was four or five years old along in there to those camp meetings. Them days they would go visiting on Friday night or Saturday, anyway, and stay over till

Monday. Then they'd swap back a visit a little later.

I taught at Mineral Point in 1898. We didn't have a closing-of-school exercise. We didn't get enough pay to have anything over there. Twenty-five dollars a month [salary].

While I was teaching in this school, there was a young man in the neighborhood. Big, gawky man about twenty-five or six, came to school barefooted to visit the school. I was busy with a class at the blackboard. He got to wiggling his big toe—he had a toe about so long—and he got to wiggling his toes at the girls. Of course, at those that was in the school, but not in the class. There was a little snickering went on, and I noticed it, so as soon as the class was over, I dismissed the school for recess a few minutes. I told them all to go out to play. Happened to be an old stove that hadn't ever been taken down, and it had a big long hickory stick laying there under the stove for a poker to turn the wood. So I went back to where this fellow was. He was still sitting there, you know, as my visitor, and I told him that it would be more pleasant in the school if he wouldn't cut up and cause the girls to pay attention to him so they could study. It was a little bit disturbing, and he got a little bit sassy. So I reached around, and I got a stick of wood, and I put him out of the house with it. I didn't hit him with it, but he went when I got that stove poker. He went on out a way. That happened on a Friday. So when I went back on a Monday, I took my cornet with me. I played in the band at this place. And also I slipped a thirty-eight in my pocket because he had a little tough name, and I was goin' to run that school when I was a-runnin' it.

The Reverend F. J. Light *(Mountain Grove, Wright County)*

I have lived here in Mountain Grove forty years, and I have pastored the church here more or less during that time. I lived at Poplar Bluff and also in Howell County around Brandsville prior to coming here.

Well, the meetings back there were very informal, of course. About all you had to do was announce your service, and they was there. People flocked to service in great crowds then. We had brush-arbor meetings and sometimes tabernacle meetings, tent

212

meetings. In this part of the country back yonder forty years ago, we had a lot of brush-arbor meetings. It was a very common thing in the summer season. In 1919 we established a Church of God campground here in Mountain Grove. We have maintained that ever since. Meetings last six or seven days now. Back in the earlier days they lasted ten days or two weeks sometimes.

When I was young, we had very little entertainment of any kind. Occasionally we would get together for a Saturday-afternoon ball game among the boys, and at school we had our literary society, which would meet together on Saturday evening, and debating society and things like that. We had some great times with those debates. Way back yonder we would debate on the Negro and the Indian, which one was the cruelest treated, and a lot of such subjects as that. Sometimes it would get pretty interesting, and sometimes a pretty hot debate, you might say. They were not particularly prepared at all. Not much preparation. They would just go in there like they was going to have an old schoolyard ball game, and they would choose sides. They'd be two that would be the main men, and they chose sides, and you spoke on the side that you was chosen on. A whole lot of people would participate in each debate. We had three or four judges that sat out, and they would give their decision. Of course, it was just all in fun whether you lost or gained.

They had kangaroo courts. We used to have them occasionally. We would meet together just anytime we got the notion to meet together, and we'd rein some feller up in kangaroo court, you know, accuse him of something and generally proved him guilty. What kind of charges? Maybe mistreating his wife or failing to water his horses. There was no fine. Maybe sometimes he would have to do something special, a little embarrassing. Religious debates, them used to be very common. I have watched them considerably, and the community was never benefited a great deal from them. They would have some very hot debates.

Literaries were well attended. The whole neighborhood would attend them. We had no other amusements. It was well attended and carried on in very good style.

213

I moved around Brownsville in 1905. I lived there for ten years. I grew up down in Arkansas, across the line. They were very peculiar in Arkansas about their entertainments. They didn't allow anything like that on Sunday. We didn't have anything a-tall. We boys couldn't get out in an open lot somewhere and just jump, like we used to have running jump and standing jump. If they caught us doing that, they would call us right out. We couldn't do it. A little later on they got to where a man would come through the country with slides or moving pictures and go to our schoolhouses and show that. That was the first entertainment that began to get into our country, in the picture line at least.

As a boy I saw nothing outside the old-time medicine men. They gave shows and sleight-of-hand performances. They would generally have a clown along with them. They sold their products.

We had box suppers. I never knew anything about pie suppers as a young lad. We had box suppers, yes.

Now entertainment in the churches, there was nothing like that among the young people. We didn't even have young people's meeting or anything like that like they do now.

In some communities it was pretty hard to have even a religious service without being disturbed more or less. Of course, I was a Church of God preacher—they called us Holiness back there—and our preaching wasn't received very good among the other churches. We got a lot of persecution. We had women preachers. They resented this very much. Some do yet, but not too many things change. Women preaching was looked upon then as being out of their place altogether.

So they would organize their forces, and where they couldn't get us out in open debate, they'd try to run us out of the country, you know. I've had eggs and rocks and frogs and I don't know what all thrown at me while I's preaching. About the funniest thing that ever happened to me was, I was preaching in a place one night. Some fellow throwed something in and hit me up the side of the head and felt wet and cold. I thought it was mud at first. I didn't pay much attention. I took my handkerchief and wiped my neck and went right ahead with my preaching, you know, to keep from disturbing the congregation. Pretty soon a couple of young ladies sittin' out in front looking down and rustling and laughing, and I looked down, and there was a couple of frogs hoppin' out there. They had hit me up the side of the head with frogs. This came in through the window. Well, I had eggs the same way. That didn't amount to very much, but our persecution in those days was pretty severe, in a lot of ways.

Joe McKinley *(Springfield, Greene County)*

Back when I was just a kid [in Wright County], we had the Friday-afternoon entertainment at the school and invited the parents in, that is, if the parents had time to make it. Everybody worked in those days and didn't have time to go, but a few would always come. Of course, when we had our spelling bees we preceded that with a program usually. Our literary society was an evening program, and children participated and then we had a regular Thanksgiving program of a religious nature, studying the things we had to be thankful for, then we had our Christmas program, and then on the patriotic side of it we would take in George Washington Day's program and those things. It finally got into kangaroo court, where the older people participated. A man would come out who hadn't shaved for a month, and they'd be judge and jury, and you'd be surprised we had always been able to get a lawyer to come out from town. In fact, he was the judge, and he'd bring someone with him that was interested in being the prosecuting attorney or something, and it was a kind of enlightening situation. The public, they really learned some things, some things that were legal and some things that were illegal, you know.

For instance, I sued a man, I remember, one time. I was a boy about seventeen or eighteen. I sued a fellow in town for alienating my girl's affection. It was a friendly affair, you know. However, I had kept company with this girl just along. We just agreeably each one went our way, and she immediately began going with this boy from the town. Well, you'd be surprised how keen that became to those people out in my community, five miles from the town where this boy lived. Boy! They went into it just like it was real. I had been imposed upon! Well, I had my attorney, and he had his attorney, and, by the way, those

215

boys made attorneys later, and it may have been that they got the incentive from that contest. The judge was a man in town, and when he made decisions, they had to go according to law. They couldn't be any just trumped-up things. They had to be, they really had to be right. We had a prosecuting attorney from town. He was really a lawyer, the judge was really a lawyer, and so we really had that phase of it, and these women would come and just sit there. They'd just act. We had our jury, and our jury voted. We would make our decision right there. I won my case, and he was supposed to pay me a penny a year. I relieved him of all responsibility since he asked for a receipt each time.

That was strictly on its own. We didn't have anything else, only the open court and proceedings. But now, of course, it took quite a little bit of planning between one month and the next. We had our witnesses, our circuit clerk, and everything that was needed. Then in between these things we would work in a ciphering bee. Some people wouldn't be interested in these unless it was between two schools. When you got down between two schools, that was a different proposition. Then we always had to have a little program about half the time. Say, meet for an hour and a half and have ciphering for an hour and a program a half an hour. And give the people a chance to hear what their kiddies could do in the way of singing or recitation or anything they did, and they performed. Now that was a kind of community proposition.

We had our debating contests too. Usually two pupils would debate two people from the community. The neighborhood would always join in with the boy, of course. Sometimes a boy would be on a panel against his father.

We had people suggest the debate topics, and then we had a committee approve them. That was how we got a library case and some library books. We had about two or three meetings leading up to that, telling what we could do, how much better off we would be in our outside reading if we had a library. Got the patrons interested, and we got a carpenter to build us a library case and, first round they bought a book for every pupil. We had about sixty, I suspect. Then we started out a book a month, let as many as wanted to buy a book a month, and that little school was the first little school in the county, possibly outside of town, to have five hundred volumes [Reed School District, Wright County, about 1896].

The schoolground in those days was donated, and it was usually the acre that the farmer couldn't farm, and neither could the school.

I remember when I first started, our people used to think you had to have poetry before you could have a recitation, and then they got to having monologues, and then it started out on dialogues, two people doing it, and then they finally got to putting on school plays. We would buy a whole school play and it would cost a quarter. And that would be a full hour or possible more play, four acts. You always had to give something that would rather fit the community, like where some crook tried to sell spurious mining stock, and then it all turned out right in the end, you know. And the teacher would pick out the talent, and we'd all work hard to get to be in that lineup, and they'd practice through the school, and they'd give it as a program at night. Never did charge. It was a free thing. People didn't have any money. We would arrange some little fund to take care of our expenses. We would have a little box supper or pie supper to get a little money. Some of those made very good. I remember one boy in our group that handled the young people afterwards for Billy Sunday [evangelist]. Reed was possibly an above-average community.

Sometimes there were two ministers at baptizings, and two were baptized at once. That was where they had their wonderful emotional interest and got lots of people in. Today it's a little different thing. I notice our preacher down here, he said, "Now if you want to be baptized kinda secret why you can. It don't have to be publicly, you know. Set your hour and come in and be baptized in the baptistry." Back in those days that was the climax of two weeks or four weeks.

About 1890 they began to be some new denominations come into this state. They hadn't time to have churches up to that time. Now the first ones was the Church of God. They had a man from Lansing, Michigan, come in here, a very able minister. I think possible he felt he was doing missionary work, and possibly he was. He came in here and got a few followers, and then

217

they'd get a campgrounds, rent tents and put up tents, start the meeting, put up the tabernacle, so they were just out rent. During that first meeting I suspect they would have one to three hundred come in, and that was more than the little churches would take in in ten years because there was an emotional side to it. They had wonderful singing, and they really had wonderful preaching. And then a little later that church split over wearing a tie. They didn't allow you to wear a tie. Women couldn't wear flowers on their hat. They were the Assembly of God people. They were possibly a little more liberal than the Church of God. Ever time we got a new denomination we had tent meetings again. They have church all the time. They have it at eight o'clock in the morning, and they have it at two o'clock in the afternoon, testimonial meeting, and that's where they get all stirred up. Then they have church again that night. Those denominations particularly had meetings. People came whether they joined or belonged or believed them. They came for the entertainment.

I remember I was going to a pie supper. The pies sold very well, and everything went off quietly till they put up the most beautiful girl for the prize. Well, after someone had had his girl friend nominated and a hundred votes voted for her at a penny a vote, then a few fellows picked the homeliest girl in the community, cross-eyes and everything, and ran her and won it. And boy, when that was over you'd better get home just as quick as you could, cause they might start something. I was at one one night where a boy was killed. One boy just shot him six times. He just shot him as long as he had any shooting in his gun. This fellow thought he had pushed him too far. That was another thing about these old times, boys back sixty years ago, they'd fight you. I know a schoolteacher out at a pie supper one night, that was at Ava, he got into it to try to maintain order. Some old boy just come at him with a knife, cut him across the nose. He had a bad scar there, but the old boy that has the scar, he won out. This fellow had said he'd never let him have a peaceable pie supper, that he was just going to break it up. He was just going to do this, that, and the other and the teacher just stayed right with it till the people got confidence enough in his ability to handle it that they would come.

218

He finally whipped the rough element out. He is a business-man today in Ava [see interview with Emmett Yoeman].

They would run Miss Black and Miss Brown. Someone would be interested in Miss Black, and they'd come around: "How much do you want to vote at a cent a vote?" Then they went up to the checkers who sat there, and if you had four dollars this trip for Miss Black, they marked her up four hundred votes. Then they put a time limit on it, and when that was up, why that was that. Why I've seen cakes bring forty dollars, and I've seen pies bring five dollars. Some fellows would decide, "Let's have a little fun tonight. Let's buy so-and-so's girl's pie if we can find out which it is," and they'd find out. They'd run it up to five dollars. The girl got the cake, and then she'd usually cut the cake and give each of her sponsors a piece of cake. It was usually eaten right there.

Fred May *(Galena, Stone County)*

They held literaries in the country schools and in the town schools. I lived in the country about a mile across the river.

They usually had heavy crowds attend court, and court usually lasted a term of a week or two weeks. There were lots of jury trials, both criminal and civil. And many times if they didn't finish the trial of a case during the day or afternoon, they'd finish it at night. Especially some of the argument of the attorneys to close the case. They had a grand jury back in them days once or twice a year. Oh, people just came to these for the entertainment, the pastime. We only had two terms a year, and that was in March and October. It was usually open weather, and people would come in especially if it was an important case like a murder case or something like that. That drew a big crowd.

The people camped down on the river. They had a campfire down on the river, about where the bridge is on the west side of the river. That was the bottom in there then. There was no settlement, no buildings, no anything, no railroad. That was just an open bottom then. They picnicked, used it for a circus ground and a general campground. Women came to court too.

Circuses came around about once a year. They came with

animals in cages, tigers, lions, panthers, and elephants.

Camp meetings and protracted meetings were pretty numerous. They were held more or less in the outlying school districts. Sometimes they would build a brush arbor and hold their camp meetings there.

Mr. and Mrs. Frank McPherson *(Ash Grove, Greene County)*

Mr. McPherson: Yeah, I lived in town since 1884. Seventy-six year; I'm eighty-six now. On Friday nights at the college here we would have our literaries. That was composed of speaking, plays, dialogues, and things in the entertainment line. The college students put it on, and every Friday night we had an open entertainment over there, and I have seen as much as three hundred people over there at night. People came from all over town, and they came in from the country too. It was Ash Grove College and School of Elocution. This entertainment was rehearsed, of course. The people from the community participated too. If they wanted a place on the program, they could get it.

Well, we used to have medicine shows here ever once in a while on the streets at night. They would have big gasoline lights, and they would have a surrey, it was good to stand up in, you know, and they would give you the medicines in front of it. They would have a colored man to play the banjo or something like that. Maybe two or three or a quartet, get some really good singin' music. Then another thing, all the churches had more entertainments then than they do now. We usually had parties, birthday parties. We had Children's Day and Boy's, Son's, Father's Day and men's meetin', men's Christian association.

Mrs. McPherson: When our class graduated, there was ten of us, and we had our graduating exercises at the old opera hall. We didn't have any place at the school building. It wasn't one or two essays then. We all had one. Had to listen to all ten of us. Each person would recite something. We wrote our own essays, our own orations. Each one was supposed to be about ten minutes long. And then, of course, between we had our solos and a short address at the closing time by the superintendent. People came to these more than they do now. They came as entertainment.

Protracted meetings, they was in their glory in those days. People really went. We didn't have a place up here, halfway to Bois D'Arc, where they could dance all night. I have seen a revival here in Ash Grove where the people stood around the windows and listened in the summertime to hear. Most of them run about two weeks. Our church, the Christian church over there, used to have a camp over at Greenfield. We would go every year. Reverend E. States, he was the state evangelist, he lived in Greenfield, and this park was at Greenfield. We used to go down there in our horses and buggy. See a whole row of them along one side of the road. Stay till after night. We had plenty of time, least ways we took it. We had extra good order. Toughs came in some times, but very, very infrequently. Hardly ever did happen.

Sam Miller *(Fordland, Webster County)*

... [I lived in] Douglas County when I was a boy. Down around Seymour. Bean hullings, and peach cuttings, and we'd always wind up by moving all the furniture out of the bedrooms and playing "Skip to My Lou" and things like that. This was a play party. My dad wouldn't let me go to a dance on a bet.

We used to have literary societies organized and run the whole winter and done some pretty doggone good things. Why I've ridden ten miles to hear some debate. Some little old debate. We'd really work on it. Which one was most poorly treated, the Negro or the Indian. There was a lot of medicine shows. They made the towns. They had some pretty hot contests, like a diamond ring for the prettiest girl. This worked as it did at pie suppers. You voted by putting up so much money, except this time the money went to the medicine show. Sometimes prizes were put up in the same way for the dirtiest man or the laziest man, etc.

A lot of them schools put in about the last six weeks getting up a program for the last day.

There was a Coty girl. She took the cake down at her school.

Then she happened to be caught over by the Reed School when they had a pie supper. And they put her up. They knew they'd get a lot of support, and she won. And it finally got to be almost a traveling pool. They'd take her, and all the people on Wolf Creek would go along, and they'd back her up. They'd save their money from one month to the next. You could get acquainted in an area of fifteen or twenty miles. People in them times was still conservative. If they spent a dime, that was about as much as they would spend, and these old boys at Wolf Creek, they would spend every dollar they had and arrange to borrow. "They's goin' to take you up to a hundret dollars for this girl."

Well, we used to have brush-arbor revivals, and we'd be down there for a month or six weeks. The Wright County people built cabins. They'd just move in. There were thirty or forty cabins around. It was a picnic and everthing combined. They had all the shouting and testimony and all that stuff. There was always somebody there that would do it. You see, they'd start it soon after lunch and run it till just in time to wash and clean up for the night services. People would come to it as they got ready. They were all camped out on the grounds. The old men would sit around and visit all day, and the women would hit it around there.

Used to have picnics all over the county. Some of them would last three days. They'd call their hogs in from the open range and allot them and sell them right there, Johnny on the spot, and drive these hogs from Gainesville to Mansfield. I've seen as many as five hundred head of hogs come right into Seymour. Put 'em on the train and ship 'em out.

I remember one time in Norwood we was having extemporaneous talks. You would just think of something and get up and talk. Sterling Williams was a freckled-faced, curly-headed fellow with a loud, raspy voice. He took "The Crooked Mouth Family" as a title, and he went through and imitated how they'd look under different circumstances, you know, and, boy, it was a hit! It kinda started him out participating in things like that because he discovered that people appreciated it. Now after that some of those pie suppers would have Sterling Williams on the program. Things like that developed.

I'd walk eight or ten miles to go to something, and when I

got there, it wouldn't amount to anything. We were hungry for something to do.

Jason Roy *(Mansfield, Wright County)*

I began my teaching career in 1905. The school entertainments might be classified as spelling bees, debating societies, or ciphering matches. I had students were able to spell every word in the spelling book at that time. I had students in my school who were very fast in the figuring, addition, subtraction, multiplication, and division, and things like that. Then we had debating societies. Different localities would choose their best debaters. They would choose a subject, and we would have a public meeting. The schoolhouse would be crowded with people. The interest became very intense. They would take a subject like "Resolved: The Negro was more cruelly treated than the Indian," and some of those old-timers were eloquent in their debates. They were extemporaneous, I would say.

I taught five terms at Brushy Knob, and I taught five terms at Shiloh, and I taught at Vera Cruz, and I taught at Cole Springs, and I taught one term down at Aurora. The terms were six months, and my first school I agreed to teach for twenty-five dollars a month. I averaged seventy-five pupils, and we taught all the grades from the first to the eighth. After the first month the school board agreed that I was worth more than twenty-five dollars a month, and they raised it to thirty dollars a month and later to thirty-five dollars a month. I taught five terms of school there in that district, and that was rather unusual. You know in those days a fellow didn't have to know very much. If he knew arithmetic and spelling, he could get along pretty good.

Almost every Friday afternoon we would give the entire afternoon from noon on till four o'clock to literary work or in spelling bees. The community would come in to this. The fathers and mothers of those boys would come in to that, and they would really enjoy it a lot. Not only that, but we had a night session, there at Brushy Knob, not every week, but about every two weeks, and that would take in the public, and we'd have visitors from some other district. Some students did pronounce the word before they would spell it, but they did not pronounce the syllable. We never practiced that in our school. But I taught rules in spelling. I also taught penmanship in school in those days. We held kangaroo courts. We did that for amusement.

In those days we had a lot of hillbilly singing and hillbilly music, and you'd be surprised in my time how much nicer and more beautiful that music was than the music we have today. We had quartets who could carry the soprano, tenor, bass, and alto, and it was really pretty to hear those quartets sing.

Oh, yes, there were some camp meetings held out under a brush arbor and some in buildings. They encamped for two or three days or a week at a time. People paid more attention to their soul's welfare in those days than they do now. The worship was considerably more noticeable in those days than it is today. I remember that I attended meeting one time in the school district where I taught. An old lady there would get happy every meeting day.

I remember that we had a fellow in our neighborhood that tried to use big words in his debates, and lots of times those big words had a far-distant meaning from what he intended for those words to mean. But we learned to recognize the sincerity of that man, and we never at any time laughed at him, because we was sincere and tried to help out. He thought that was quite a brightness to utter big words. You'd be surprised what good order we had, no toughs or drunks.

I remember one time when three fellows came in from another neighborhood and they kept running in and out the door and making a to-do about it. One fellow would step just as high as he could step and let his feet down on the floor. I went to him and told him, "Now you're just as welcome here as the flowers in May, but we want you to behave yourself and don't go outdoors any more now, stay right here." When our program was over, he and I started out the door. He said, "Watch out, boys, here comes the governor! He'll make you behave yourself." And I walked up to him and I said, "Are you going to behave yourself or aren't you?" He ripped out an oath and told them what he might do, but he hadn't more than said that till I knocked him about fifteen feet right down the hill. He didn't bother any more. That was our last trouble. That took care of that.

In those days Thanksgiving was one of our great jubilations. All the women and all the men, all the people in the neighborhood would bring out well-filled baskets, and some would bring wild game. Some would bring wild turkeys or homegrown chickens. Those women were great hands to cook, and we had programs on Thanksgiving that were in accord with the Thanksgiving season. Oftentimes, also, we had Christmas programs and small gifts for the students, and then on the last day, we would have another great gathering. They would bring in well-filled baskets, and all the people of the district would come in, and the children would have a program of recitations, or "declamations," we would call them, and the parents would make long-drawn-out speeches, lots of them. We really had a great time on the last day of school.

No, no we never put on any plays. Nothing more than just short dialogues. We had dialogues in our entertainments, but we never had a regular drama, regular play. We never did tackle that in our school. We had nice little things made of three or four several different acts. I had addresses of people who put those things [dialogues] out, and I would send and get maybe four or five, and I would pick out the ones that were best suited for our talents. We'd have music in those days second to none. We had lots of good fiddle players and guitar pickers, and they were really fine. I don't think you should forget about the old-time square dances.

The home talent and the home people in those days would go further and make greater sacrifices—that is, they would go further in proportion to the means of travel. They only had wagons and horseback, but I believe the entertainment that the people had fifty years ago and sixty years ago—and I can well remember those modes of entertainment—I think the people enjoyed the entertainment then much better than they do now and even meetings and Sunday school and church of any kind. I believe it was more enjoyable, that the people got more enjoyment out of it than they do now. I think the affection and the love that one neighbor had for another was greater in those days. We've lost something in this modern civilization.

I remember that the teachers in those days on the last day of school had to do something modest, to bring some candy to treat the children, and I remember that one teacher brought out half of a small poke of candy and gave a stick apiece to a woman and four or five of the children and he told the children he said, "That's all the candy we have. We were short on money and didn't have any money to buy any more with." And of course, our spirits all sank when he told us that. Then in a little while he reached up in the attic and brought out a big bag, a big sack of candy, and we were very happy that day.

The discipline in the schools in those days was much different in those days than it is now. I started out in my teaching thinking that I had to use corporal punishment, but I soon dispensed with that. I found out I could do a great deal more with boys and girls by talking to them than I could by using that hickory.

The kangaroo courts were handled just as much like a real court proceedings as we could very well handle. We would charge a man with stealing a pig or riding a man's horse away or turning a man's horse loose at church or something like that. I remember one time we charged a fellow with beating another fellow's time with a girl, and that was funny. I think he finally married her too. Well, we couldn't very well fine, but we'd make a ruling that they would be unable to exercise or to be in the company of all good men for a period of thirty days.

At the Oak Hill schoolhouse we had a revival there in the fall of ninety-three. They had a big revival that lasted six or eight weeks. There were a lot of people converted there. I can remember mighty well the preachers. The old-fashioned revival preaching was what they call now fundamental principles. There was more arm waving. People liked this. It was what they expected.

Occasionally we would have literary meetings, but they didn't last long. At that time the boys in the community didn't get along too well. There would be fights and disturbance, and usually it didn't last long. They would get in trouble, and they would close it.

The first traveling show that ever I knew come through this place in 1903. A man and a woman put on a little show. It was just a little old kind of a dialogue of a thing.

Our literaries would be debates.

Medicine shows, they didn't come in here very much till

after 1910. I don't remember any medicine show till after 1910. Wasn't enough people to come around, I reckon.

L. O. Wallis *(Ebenezer, Greene County)*

We kind of made our own entertainment. We used to have school entertainments. We spent a good deal of our time practicing up, and we would go all over the country to the different little schoolhouses. We went as high as five, six, or ten miles around to all the school entertainments and all such as that. They'd be plays, "dialogues," as we called them, and so on. Of course, some speeches and some musical programs. Usually someone in the community could play the violin or some other musical instrument.

We had some debating societies, but I never attended too many of those. There was a permanent shed on the camp-meeting grounds, and they held meetings there on up until 1901 or 1902, along there anyhow. There was virgin timber all around, and the church owned twenty acres of land at that time. This was church land, Methodist, South Methodist, it was then. You see, there was the South and the North. They merged several years ago, but that was the South Methodist Church. People came from miles around and camped there for a week or two weeks. Have their tents and one thing and another. It was a permanent shed, wasn't boxed up or anything but had a shingle roof and was built good and strong. Oh, they had a few benches at the back, but most of them was just about a two by twelve, boards, you know, up about so high, and the meetings were always held about this time of the year (mid-August) after the people had thrashed, and some of them would go out to their straw pile and haul in several loads of straw and have it about, about that deep, with straw all over the floor of the shed. It was a dirt floor. We had a little platform about eight by ten where the preacher could stand, and there were a few chairs set on that for the speakers that were special, and the organ was down on the ground. They'd haul it over from the church.

A number of people would come to these who were not Methodists. People would be traveling through the country and hear about it and stop sometimes. I remember a young minister that happened to be coming through here. I don't know how he happened to be coming through here. He was from the southern counties, and he stopped and made several talks to the meeting there. He was just a lad about seventeen or eighteen years old. He went on and made a Methodist minister.

We used to attend these meetings, and one year we camped. From 1897 to 1901 we lived five miles north of Ebenezer here. And one of those years I remember particularly my folks camped here. About '99 or 1900, I don't remember which, we camped. An old colored fellow worked for us, Old Bob Small. Old Bob grew up a-workin' for my father, and so we left him to take care of things and rode the old saddle pony up here and left him up at my grandmother's. And so one afternoon along about the middle of the week somewhere, my father sent me back down to see how things was getting along. It come up a rain along the middle of the afternoon, and we tied that old pony up in the runway of the barn behind the wagon there, and we went up to Old Bob's room and laid down and went to sleep and slept all night. And the next morning about daylight we woke up, and Bob woke me up and said, "Here, boy. You know it's come daylight?" He said, "You get your pony and get back to your Daddy."

We had a tent, and we would cook on an open fire. We didn't have gas stoves and things of that kind then, you know. People would bring in their supplies with them. A lot of them came in a covered wagon and would camp out of the wagon. They would go out here to a farm nearby and get a few bundles of oats, enough for the horses.

The campers started. They usually had a sunrise prayer meeting. And then they'd have afternoon services and night services. They'd have maybe, before the night service back in those times, they would have a special prayer meeting. A little group would get off out away from the shed, away from everybody, and have a little prayer meeting. I can remember those quite well.

I remember one instance in particular. A fellow by the name of Balley, I believe—could be mistaken about the minister's name—he was preaching, of course—it was special revival services—and he was preaching on the heart, what that meant to people from a religious standpoint and so forth, and he had a glass of water there and a little stone in it with a string tied around it, and he give the illustration when the change came and so on, when we professed religion, why, he lifted that stone out of the water, and that was getting the stone out of our heart. Made an impression on me. I never forgot that.

The style of preaching then was quite different. They really preached hellfire and brimstone. You don't hear much of that any more. There was no compromise in their preaching at that time. They didn't pull any punches. One thing that you never see any more that was very outstanding then was a lot of people shouted. I've seen fifteen or twenty people shouting at once. Sometimes during the preaching or—maybe not during the preaching—during the service following the preaching—when they'd call for penitence and so on, or maybe sometimes it would pretty near break up the service there for a little while during the preaching. If there was a real good preacher with a strong appeal, and then they used to have testimony services almost every day. A lot of people would tell their experience and what the Lord had meant to them. That was what we called the testimony service. That was one of the outstanding features that I recall.

I never attended a church trial, but I have the old church records here where they held the old church trials, and they suspended a member for a period of time. I forget what the offense was, but it's in the old church records here. The church here was organized about 1832 or '34.

The pastor of the church would be here, and we would have a special evangelist come in and, sometimes there would be two or three or four that would drop in for a day or two at a time, but there was usually one man who came and stayed through the two weeks' period. No groups other than the Methodist used this as groups. Of course, the shed was used. We used to have Sunday-school picnics, the Fourth of July, and all the Sunday schools in the township were invited to participate. Each one of them brought part of the program, and we had a big dinner and so on, and people were allowed then to put up stands, and that was just a regular Fourth of July picnic. We had stands for lemonade, soda pop, and firecrackers. That was before the day of much ice cream. Ice cream was a rarity back in those days. We would have readings and songs. Each Sunday school would have their part of the program, religious, patriotic. And have some speakers, have a minister or two to speak. I remember at one of those when I was just a little shaver there was a speaker by the name of Swanson. He was pastor, I think, at that time of Fair Grove Methodist Church, but he made a talk to us youngsters, and he wore to the Fourth of July a vest, and when he got it buttoned up he had started one hole wrong, and he said, "Do you see what I have done here? I started that bottom button at the wrong place, and ever one of these are out of place." And he give an illustration. He said, "Now if your lives are started wrong and never corrected, why as long as you go along you are getting just a little farther wrong all the time." He says, "Now all these buttons are wrong because I started out one wrong." He made that illustration.

I attended my first real court that I remember of in the old court house on College Street there. A man had killed a woman out about Bois D'Arc or somewhere in there. I think I was about twelve or fourteen years old. Old Tom Delaney was the defense attorney. I remember that quite well.

They picked up their jury out on the street instead of having them chosen as they are now. They would just go out on the street, and there was some fellows that spent most of their time during the court day expecting to get on the jury. Well, they tell this on Old Joe Pollack, and I expect it's true. Well, Tom Delaney and—I can't think of the other attorney, he was a fine man and an outstanding speaker. Anyhow, they were opponents in a case, and old Joe always laid around there, and a lot of those fellows had the habit of borrowing a little money off of a fellow you know and kinda promise to support him. Mr. Pace was a neighbor down north of here, was on the jury at the same time, and he told us an actual happening. He said they were waiting there one morning for court to convene, and this other attorney said, "Well," said he had one juror on his side this morning, said he just loaned Old Joe Pollack two fifty. Said Tom Delaney popped his head down on the desk, said, "I just loaned him five!" So I guess that was kind of a common practice in those days.

People came to the meetings from all around the community. The whole woods would be tied full of buggies and horses. We had a little trouble with rowdies once in a while, but the substantial people, the men, usually took care of that all right. I don't remember there ever being an arrest made. One night I do remember hearing them tell about back before my time. You have seen pictures of the old freight wagons where they had the lock chains on them around the wheel going down the hills. Well there was a freighter came by, and I think my grandfather or grandmother saw that he came by and camped there for a day or two and had his horses tied right around the wagon, had them tied to the wagon, and came up a severe thunder storm and killed all those horses, and this man carried a feather bed with him, and he was in the wagon and lying on this feather bed. Killed those horses and melted that lock chain—the links melted together—didn't kill him. You have always heard that feathers give you some protection against lightning, and that ought to be a pretty good instance.

Charlie Weaver *(Springfield, Greene County)*

We was always just kinda entertainers from the start when we was just kids. My father was a banjo player ahead of us, and we kinda taken up the music end of it. From that night when I was just about eight years old, I played an old Autoharp. 'Tain't a zither. It's an Autoharp, and my brother [Leon] played the banjo. In later years he picked at the mandolin. Well, we'd go to parties, and at these parties they danced, but they'd sing when they danced. They didn't have music, they'd sing their songs. Then they had the old-time fiddlers. These old-time fiddlers they'd have they called them "break-down fiddlers" and stuff like that. That was taken from the Irish. They taken that from the Irish from the Irish reels. The first entertaining I did anywhere, that was literary. I'll never forget the night. A cousin of ours was a schoolteacher, and we'd play around home, and he kept wanting us to come to Ireland. In them days they never had no entertaining to speak of. All they had was a-singin' and like that. So he had us come own, and, of course, we blacked up, me and my brother did, and comin' home we walked a mile and a mile and half to the school house, and comin' back through

231

the orchard a big old possum run out in front of us, and we grabbed that possum by the tail and was carryin' it home, you know, and the way we was blacked up like we was a kind of niggers, and I always thought of that.

Musicians was just as scarce as hens' teeth. Oh, you'd find a fiddler once in a while or a banjo player, but you'd hardly ever find a guitar player at all. So then I taken up guitar. Well, brother Leon he finally got a-hold of an old mandolin, one of these that looked like a gourd, and he got to playing a mandolin, and he got pretty good on it, and we got to where we could memorize. And 'long about 1908 musical comedies come through like *George White's Scandals* and *The Merry Widow* and all that stuff like that. We memorized all of those selections. We saw them in Springfield. We went to work at the old Baldwin Theater. The first entertaining we done was medicine shows.

You see, a medicine show come to town, and he would have some blackface or something getup, you know, and sing a song or do a dance, you know, to hold the crowd. Now Odd [O. O.] McIntyre, out of New York, was a great critic, used to write for *The New York Times*. He said that the medicine-show man had to have more than the man on the stage because the people could walk away any time they wanted to, and the people had to set in the theatre. We went to playing towns for a medicine show and follered that for some time, and then we got to going to all different kinds of stuff. And finally my brother—I got tired of traveling around over the country and makin' these fairs and stuff like that, and so I quit the road—and he went on and picked up a fellow that was with him, my older brother did. He works on about two years and got on the Pantages Walkin' Circuit, and he was on there a couple of years and come back and picked up me and my other brother, Frank, and made our show business. We was working for the Keith and the Orpheum. We have knew what it was to work for a small-time outfit. We worked for the Keith and the Orpheum, the best then, and played at the Palace in New York. It was the dream house of the world then. There were thousands of people who worked up to play the Palace and never did make it. We played all the big cities in the United States.

Now back in the days when I was a boy, people would all

meet, and they'd have a party and popcorn, molasses candy, and they would have a "taffy pullin'," they'd call it.

These women used to dress in calico. They didn't have these fine dresses and stuff. They would always get these loud colors on the Fourth of July, and then it would come a rain, and it would all fade. Boy, they'd look a sight.

As fer as foxes, I've been a hunter ever since I was a kid ten years old, and I was huntin' before then, but I was actually a hunter when I was ten years old. I know my folks didn't want me to keep dogs, and I'd keep them in a pasture away off away from the house. They used to have entertainments at fox hunts then. We used to have these hunts, and back in the early days you take—the old fellows would drink, but didn't drink to excess. They'd take a drink or two and set down and talk, and you'd never know it, but a lot of younger fellows, they'd go out and they'd get drunk. Well, we quit havin' these hunts, big gatherin's, because somebody would come in there drunk. Well, so I started—when J. W. Crank died—why I started handlin' the Greene and Taney County Hunt, and so I told the boys, I said, "They ain't but one way we're goin' to do this. The only way you can do it," I says, "to stop the drinkin', clean it up to where your mother, your sister, or anybody else can come here and be respected." And I says, "We can get back on our feet." Well, of course, one guy—he was a smart aleck—he said—I was talking, and I was pretty wild when I was younger myself—and he says, "What are you goin' to do?" And I says, "I'm goin' to clean up my door and then I'm goin' to yours." And that's what I had to do. So this preacher, Swatney, over there, we got him, and we had him every year over at this hunt, and now they's three or four preachers goes to these hunts. Now, you don't hardly ever see a hunt that they don't open it with a prayer and close it with a prayer. At a hunt through the day they'd just visit around and swap dogs and stuff like that and have some entertainment. But the biggest part of the hunts goes on at night. Of course, some hunts took up a little different at different places. Worlds and worlds carries a card in the hunt and never had a hound. They like it because the people comes there, and they bring an entertainment. We hire the best acts we can find. That's where Slim Wilson and some of them got their start. We liked to have

233

some good singin'. We like to have a choir singin' and stuff like
that. And that's where we come on top. We got the people outside
the hunt interested. A hunt generally goes on three to four days.
We have field drives, and when they have field drives it generally
goes on about three days.

We originated handsaw music. They was the first handsaws
ever played on the stage. We made that one string fiddle thing
and played it with a derby hat.

The old square dances at the houses, they'd have a wood
cuttin', and then that night they'd have a dance, pop corn, pull
candy, and stuff like that for entertainment, and that's what they
do when the women would be a-quiltin' and the men would be
out, maybe havin' a rail splittin'. Different things that a fellow
had to do, and they'd help each neighbor out that way through
the fall.

We'd play music for ever kind of thing, I guess, there is in
the world. Prizefights and fairs, and one time we played when
Anheuser-Busch had a big blow-out on the square, had a cen-
tennial-like thing on the square.

In medicine-show days, my job of a morning was to get up
and get a big washtub and stir up some kind of yellow powder
in gasoline, and we'd bottle that and label it "corn medicine,"
and that's what we'd sell, and we sold a medicine there as a tonic
made in Wichita. "Sultana" was the name of it. He was makin'
a pitch one night, and he made this remark: "Well, now, I'll tell
you," he said, "this medicine takes the place of all this other
stuff." And he said, "Well, now, here you take a disease like these
people say they got appendicitis. And who the hell ever heard of
a nigger bein' operated on for appendicitis?" And they'd have
some fellow come in there and say, "Anybody in my hearin' that's
got a corn or bunion, why just walk up here to the platform, and
I'll stop the hurtin' and it won't cost you nothin'." Well, he'd
take this gasoline and a sponge, and he'd say, "Don't take your
shoe off. Just leave your shoe on. Now it won't bother your shoe
at all, it won't hurt it." He takes that sponge and dip it down
there and put it on that and get it pretty damp. Well, that gasoline would just go through there like that, and that old corn
would just ease up and feel cool. Whish, I just sold that can after
can like that. A dollar a bottle, and in them times a dollar was

worth something, too. You could see a pretty good show at a medicine show. A lot of those entertainments went on to big time and made good.

One fellow, we went into a little town, and he had been under the care of doctors for a year or two. He was a pretty well fixed fellow, and some of these fellows went and said, "I'll tell you Doc, if you can do anything for this old man up in the hollow here, if you can do anything for him, you can sell all the medicine you can bring down here. That old man has been sick a couple of years and went to all these different doctors and done no good." The doc went up and looked at him and, hell, he wasn't no more a doctor than I was as far as that was concerned, and he came back and fixed up some croton oil and some stuff and went up and give the old man and worked him all out good. And the old man come down to the show the next night and hadn't been out of the house for a couple of years. He walked up to the platform and told the people, made a statement, told the people, "This man here knows exactly what he's doin'." And he didn't know no more about medicine than I do.

I've slept a many a time with a tent peg under my pillow so that if anybody would come in from under the tent there, I could take care of him. Man, it was mean and rough, man! You had to be an outlaw to even get along in that part of the country.

We set up our own tent, carried everthing right with us. We never played in opera houses as medicine show. We just had a big platform and our lights, and them was old gas lights. We didn't have no electricity. We would entertain a while and then stop and sell medicine and then entertain some more. We didn't play the Ozarks much with medicine shows. You had to go where there was lots of people working. We had a pretty big show at that. There was Leon and I, and there would always be two or three other performers and the doctor and an assistant, and then there'd be his wife, or something, or some other lady. By God, they put on some real shows, some of them.

Leon and I came up in this world. We started at the old country literary in Elk Valley. That was the first time we was ever before an audience. I was ten years old. That was about '95.

Getting back to the older days. You didn't see people as

often because there was no way of getting around. Transportation then was nothing but the horse and buggy. Talking about the automobiles, there was an old man, a Dutchman, somewhere from Germany, lived there by Ozark. Name was Joe King, and it was about six mile, and he had one big old yellow horse, and he had a wagon. He had two horses, and one of them died, so when he come to Ozark he'd hold up one side of the neck yoke and the horse would work on one side and he on the other.

Emmett Yoeman *(Ava, Douglas County)*

In ciphering matches they would choose up sides, and then they would bring up two at a time, and they would usually draw for their choice then, whether it would be addition, subtraction, multiplication—usually it would be addition. Then another pair would come up, and it was kept track of on a point basis.

The whole community went to the End of School. It was a community affair. I recall one year when I was teaching we had four school districts get together. I had my group there, about fifty or sixty. We loaded them up in spring wagons and hauled them over there about eight miles. Had a big basket dinner, and each school had then a part of the program. We had what they call a "literary program." Each teacher determined what type of program he would have. This program definitely had a bearing on whether a teacher was invited back the next year. Because the community spirit then was much stronger than it is now. They were really community conscious and if someone didn't cooperate, didn't fulfill his part of it, he just didn't quite measure up. It didn't actually represent the work done at the school, but as far as the people was concerned, it was a comparative proposition. If a fellow showed up real well, they'd say, "He's doin' a good job over there," and if he didn't, "Something wrong over there at Walnut Grove. He didn't do much." So each one did his darndest to put on a good program, one that would entertain the people.

The school in which I taught, two years before I was down there, had been shot up. I mean literally shot up. Roughnecks come through there, and they just shot holes through the building, and ran the teacher off. He never did come back. But our

236

school wasn't a rough school there. It was all right. But Lone Star, now that was a rough spot [said as a joke, from tone of voice].

Every district tried in some way to raise a little extra money. Of course, tax rates were low. Taxes furnished nothing in an incidental fund for supplying needs there for maps or globes or library or the extra things that were necessary for good teaching. So the more progressive teachers would have pie suppers or box suppers, with a program, of course, to try to raise some money for books or extras that would help them in their work. That was what I was doing one night, a box supper, when a row started. Everthing had been going along pretty well. Mr. Capers was selling the boxes for us, and we had already given our program, and I heard a disturbance in the back—rather, it was outside—and I just had stepped back there to see what the trouble was. I thought it might be some of my boys back there. And this big drunk in there—and as I opened the door there to see what was going on, he just hit me. I didn't realize at the time I was hurt at all. It stung like a slap, and in fact that's what I thought he had done, slapped me, I didn't realize that he had the knife, and, of course, I wasn't so big, but I did knock him down, but as he fell, he clinched me, and that's when he got me on the neck and on the back up here. Then I knew he had a knife, so I grabbed his arm—almost took my thumb off—I grabbed the knife instead of the arm. It was pretty rough. I would say this rough stuff was a throwback to the pioneer outfit that came in here. They were mountain folks. They had grown up in an area where moonshine and bootleggin', beatin' the revenuer, was everyday, and, of course, incidentally, they carried it along with them. I don't think a great deal of it was intentional. I think a great deal of it was incidental. They got a little more of their rotgut liquor than they could handle and it tended to bring out a little of the mean in them.

There was in each community probably, but I would say a minority, those that were a constant disturbance. Now, I know that in the community where I grew up they used to have a midnight dance. It was a common practice. It would just be the neighbors there, and they were always well conducted. But frequently there would be a group, maybe just four or five, that

237

would come in there, and it was their delight if they could break one up. Frequently they were hauled home in a wagon. Someone would knock them in the head. Didn't kill him, but that took care of him. Incidentally, that fellow was killed up here on this corner two weeks later when a fellow shot him. There was a throwback there.

Revivals? I think "recreation" would be a pretty good term to describe them. It was not nearly so much spiritual as it was recreational. It was given over to a lot of loud preaching without much thought, long talks without any planning, and lots of singing and lots of shouting, lots of testimonial, and it was a long-drawn-out affair, lasting maybe two or three hours. I would say that it was as much a matter of entertainment and was devised for that purpose more so than any spiritual purpose. I knew some of those old preachers of my own knowledge that just preached like fury there and prayed like fury and then got out and drink all the liquor they could get a-hold of. That had not a thing to do with their preachin' on Sunday. That was a different thing entirely.

I can recall when I was a boy here. That was back in the days of the horse and buggy and wagon, and people would just camp for blocks and blocks around here and come and stay all week, they didn't have any business except just to sit in up here and listen to the sensational trials. Now especially if we happened to have a murder trial or a rape case or something of a sensational nature.

. . . I can remember one other feature about it [the hanging of Ed Perry]. The old courthouse set out here in the square, and in one end of the corridor there he was laid out there, and they let the people go by and look at him.

There was an old three-story hotel out here, and people were just all over it, and one fellow in the excitement fell off it. He was drunk, and he lit on a pile of loose shingles down there, and it didn't hurt him.

People came for the thrill of the thing and the excitement attached to it. A morbid entertainment, and the same thing for the people who came here for our court trials. It was not that they had any particular interest in it, but they didn't have any type of entertainment that we have today, and so they had to devise something.

Now up until I was a grown young man, we very patriotically celebrated Fourth of July. And we would have a parade, and we would have all the states represented, and we would have Miss Columbia all dressed up there. We would go down there, and we would have someone who read the Declaration of Independence, and we would set back and listen to it. It was usually the best reader in the community. Then we wound up with a patriotic oration and a basket dinner. Then it was just open entertainment, just whatever we might have there. Usually fireworks at night.

Notes

CHAPTER 1

1. Hugh Park, ed., *Schoolcraft in the Ozarks;* reprinted from Henry Schoolcraft, *Journal of a Tour into the Interior of Missouri and Arkansas.*
2. Robert Flanders, personal communication.
3. *Ash Grove Commonwealth,* October 7, 1886.
4. *Marshfield Chronicle,* October 10, 1895.
5. *Ash Grove Commonwealth,* September 28, 1905.
6. Carl O. Sauer, *The Geography of the Ozark Highland of Missouri,* p. 223.
7. *Houston Herald,* May 18, 1905.
8. *Houston Herald,* March 6, 1902.
9. *Ash Grove Commonwealth,* November 26, 1896.
10. *Cassville Republican,* March 3, 1898.
11. Walter O. Cralle, "Social Change and Isolation in the Ozark Region of Missouri" (Ph.D. diss., University of Minnesota, 1934), p. 81.

CHAPTER 2

1. *West Plains Journal,* January 19, 1899.
2. *Ash Grove Commonwealth,* March 28, 1889.
3. *Douglas County Herald,* December 10, 1891.
4. *Cassville Republican,* July 15, 1897.
5. *Ash Grove Commonwealth,* January 6, 1887.
6. *Taney County Republican,* November 21, 1901.

7. *The Houston Herald,* February 9, 1899.
8. *Ash Grove Commonwealth,* March 10, 1892.
9. *Ash Grove Commonwealth,* December 27, 1906.
10. *Ozark County News,* October 20, 1898.
11. *Cassville Republican,* December 17, 1896.
12. Vance Randolph, *Ozark Mountain Folks,* p. 25.
13. Wayman Hogue, *Back Yonder,* p. 196.
14. *Cassville Republican,* October 7, 1897.
15. *Cassville Republican,* December 17, 1896.
16. *Ash Grove Commonwealth,* February 14, 1907.
17. Randolph, *Ozark Mountain Folks,* p. 27.
18. *Ash Grove Commonwealth,* February 7, 1889.
19. *Cassville Republican,* March 13, 1902.
20. *Ash Grove Commonwealth,* December 13, 1894.
21. *Houston Herald,* September 21, 1899.
22. *Ash Grove Commonwealth,* January 28, 1897.
23. Ibid.
24. *Ash Grove Commonwealth,* December 16, 1897.
25. *Lawrence Chieftain,* March 17, 1887.
26. *Cassville Republican,* May 23, 1895.
27. *Ash Grove Commonwealth,* May 18, 1893.
28. *Ash Grove Commonwealth,* January 29, 1891.
29. *Ash Grove Commonwealth,* March 14, 1889.
30. *Lawrence Chieftain,* March 11, 1897.
31. *Ash Grove Commonwealth,* February 8, 1900.
32. *West Plains Journal,* July 28, 1910.
33. *Houston Herald,* October 16, 1902.
34. *Lawrence Chieftain,* February 11, 1892.
35. *Stone County Oracle,* February 26, 1903.
36. *Taney County Republican,* June 4, 1896.

CHAPTER 3

1. *Lawrence Chieftain,* June 24, 1897.
2. *Lawrence Chieftain,* May 6, 1897.
3. *Lawrence Chieftain,* June 24, 1897.
4. *Lawrence Chieftain,* February 15, 1888.
5. *Cassville Republican,* February 28, 1895.
6. *Taney County Republican,* January 16, 1896.
7. *Cassville Republican,* April 18, 1897.
8. *Ash Grove Commonwealth,* March 13, 1890.
9. *Lawrence Chieftain,* April 27, 1893.

10. Guy Howard, *Walkin' Preacher of the Ozarks*, p. 165.
11. *Houston Herald*, February 6, 1902.
12. *Ash Grove Commonwealth*, April 23, 1903.
13. *Lawrence Chieftain*, February 23, 1893.
14. *Ash Grove Commonwealth*, April 24, 1890.
15. *Lawrence Chieftain*, April 27, 1893.
16. *Douglas County Herald*, March 8, 1894.
17. *Cassville Republican*, December 28, 1895.
18. *Lawrence Chieftain*, March 1, 1888.
19. *West Plains Journal*, June 18, 1896.
20. *Cassville Republican*, December 28, 1899.
21. *Christian County Republican*, January 26, 1899.
22. Edith Packer, unpublished diary, 1913.
23. Howard, *Walkin' Preacher of the Ozarks*, p. 165.
24. *Douglas County Herald*, March 8, 1894.
25. *Taney County Republican*, March 19, 1896.
26. *Houston Herald*, December 29, 1898.
27. *Ash Grove Commonwealth*, May 9, 1899.
28. *Ash Grove Commonwealth*, April 11, 1895.
29. *Stone County Oracle*, February 26, 1903.
30. *Ash Grove Commonwealth*, April 26, 1900.
31. Mrs. J. W. Shoemaker, *Delsartean Pantomimes*, p. 14.
32. Ibid., p. 65.
33. *Ozark County News*, February 20, 1896.
34. *Ash Grove Commonwealth*, April 27, 1899.
35. *Ozark County News*, May 26, 1898.
36. *Ozark County News*, February 17, 1898.
37. *Houston Herald*, December 29, 1898.
38. *Houston Herald*, February 19, 1903.
39. *Marshfield Chronicle*, January 19, 1899.

CHAPTER 4

1. *Ash Grove Commonwealth*, September 27, 1888.
2. *Douglas County Herald*, March 30, 1893.
3. *West Plains Journal*, February 11, 1897.
4. Interview with Mrs. Eva Dunlap, Ash Grove, Missouri.
5. *Cassville Republican*, February 20, 1894.
6. *Houston Herald*, February 13, 1902.
7. *Houston Herald*, January 31, 1901.
8. *Houston Herald*, September 17, 1908.
9. Guy Howard, Ibid.

10. *Ash Grove Commonwealth,* August 27, 1891.

11. *Houston Herald,* March 6, 1902.

12. *Ozark County Times,* April 23, 1891.

13. *Houston Herald,* September 26, 1901.

14. Frances Lee McCurdy, "Orators of the Pioneer Period of Missouri" (Ph.D. diss., University of Missouri, 1957), p. 136.

15. *Ash Grove Commonwealth,* August 21, 1890.

16. *Christian County Republican,* July 28, 1904.

17. *Houston Herald,* December 29, 1898.

18. *Cassville Republican,* May 27, 1897.

19. *Cassville Republican,* February 23, 1899.

20. *Cassville Republican,* July 22, 1897.

21. *Cassville Republican,* September 7, 1899.

22. *Ash Grove Commonwealth,* October 18, 1894.

23. Otto Earnest Rayburn, "Ozark Folk Encyclopedia," vol. R6, "Religion."

24. *Ash Grove Commonwealth,* May 14, 1909.

25. *Ash Grove Commonwealth,* September 22, 1887.

26. *Ozark County News,* December 7, 1893.

27. *Lawrence Chieftain,* March 17, 1887.

28. State Historical Society of Missouri, "One Way Ticket to Pearly Gates," *Douglas County Herald,* June 9, 1949.

29. Wiley Britton, *Pioneer Life in Southwest Missouri,* 9:143.

30. *Lawrence Chieftain,* August 21, 1890.

31. *Douglas County Herald,* September 1, 1898.

32. *Douglas County Herald,* July 25, 1889.

33. *Cassville Republican,* August 13, 1896.

34. *Ash Grove Commonwealth,* October 21, 1886.

35. Barnard A. Weisberger, *They Gathered at the River,* p. 235.

36. Joseph Nelson, *Backwoods Teacher,* p. 148.

37. Weisberger, *They Gathered at the River,* p. 236.

38. *Lawrence Chieftain,* June 17, 1897.

39. *West Plains Journal,* April 21, 1898.

40. *Marshfield Chronicle,* December 17, 1896.

41. *Cassville Republican,* August 5, 1897.

42. *Christian County Republican,* September 9, 1901.

43. *Ash Grove Commonwealth,* April 21, 1887.

44. *Ash Grove Commonwealth,* September 5, 1895.

45. *Douglas County Herald,* July 11, 1889.

46. *Ash Grove Commonwealth,* June 16, 1892.

47. *Lawrence Chieftain,* May 31, 1894.

48. *Douglas County Herald,* July 11, 1889.

49. *Marshfield Chronicle,* May 23, 1889.

50. *Christian County Republican,* July 6, 1899.
51. *West Plains Journal,* July 18, 1897.
52. *West Plains Journal,* June 30, 1898.
53. *Lawrence Chieftain,* May 31, 1894.
54. *Marshfield Chronicle,* June 16, 1892.
55. *Ozark County News,* June 30, 1904.
56. *Ash Grove Commonwealth,* July 24, 1890. The site of the contest changed with each question, presumably so that each contestant could appear in his own territory, in a sort of home-and-home arrangement.
57. *Lawrence Chieftain,* July 29, 1886.
58. *Cassville Republican,* February 7, 1895.
59. *Ash Grove Commonwealth,* December 3, 1896.
60. *Ash Grove Commonwealth,* May 28, 1891.
61. *Taney County Republican,* June 18, 1896.
62. *Douglas County Herald,* February 16, 1893.
63. *Houston Herald,* March 2, 1899.
64. *Houston Herald,* August 2, 1906.
65. Rayburn.
66. *Lawrence Chieftain,* September 23, 1886.
67. Rayburn, "Ozark Folk Encyclopedia," vol. R6, "Religion."
68. *Ash Grove Commonwealth,* August 23, 1894.
69. *Lawrence Chieftain,* March 30, 1893.
70. *Houston Herald,* January 21, 1901.
71. William Neville Collier, "Ozark and Vicinity in the 19th Century," p. 23.

CHAPTER 5

1. *Taney County Republican,* December 23, 1897.
2. *Cassville Republican,* February 17, 1898.
3. *Douglas County Herald,* March 17, 1887.
4. *Houston Herald,* June 18, 1903.
5. *Douglas County Herald,* December 24, 1891.
6. *Taney County Republican,* April 6, 1899.
7. *Christian County Republican,* April 2, 1903.
8. *Ash Grove Commonwealth,* June 2, 1892.
9. *Cassville Republican,* February 25, 1897.
10. *Ash Grove Commonwealth,* April 4, 1895.
11. *Ash Grove Commonwealth,* February 21, 1895.
12. *Ash Grove Commonwealth,* October 23, 1890.
13. *Marshfield Chronicle,* December 22, 1898.
14. *Ash Grove Commonwealth,* June 2, 1892.

15. *Cassville Republican*, March 3, 1898.
16. *Cassville Republican*, December 29, 1898; January 5, 1899.
17. *Cassville Republican*, January 17, 1895; January 31, 1895; February 7, 1895; February 14, 1895; February 21, 1895; February 28, 1895; March 7, 1895.
18. *Ash Grove Commonwealth*, September 5, 1895.
19. *Douglas County Herald*, August 15, 1895.
20. *West Plains Journal*, August 27, 1896.
21. *Cassville Republican*, April 21, 1898.

CHAPTER 6

1. *Douglas County Herald*, December 19, 1907.
2. *Douglas County Herald*, November 18, 1909.
3. *West Plains Journal*, March 23, 1899.
4. *Stone County Oracle*, October 8, 1903.
5. *Houston Herald*, December 5, 1901.
6. *Houston Herald*, December 5, 1901.
7. *West Plains Journal*, January 10, 1910.
8. *Ash Grove Commonwealth*, December 5, 1907.

CHAPTER 7

1. *Stone County Oracle*, June 25, 1903.
2. *Houston Herald*, July 11, 1901.
3. *Ash Grove Commonwealth*, July 11, 1887.
4. *Ozark County News*, May 28, 1891.
5. *Ash Grove Commonwealth*, July 9, 1903.
6. *Houston Herald*, June 26, 1902.
7. *Taney County Republican*, July 5, 1900.
8. *Stone County Oracle*, July 7, 1904.
9. *Taney County Republican*, June 13, 1901.
10. *Cassville Republican*, July 8, 1897.
11. *Lawrence Chieftain*, July 11, 1889.
12. *Cassville Republican*, July 8, 1897.
13. *Cassville Republican*, July 11, 1895.
14. *Ash Grove Commonwealth*, July 7, 1898.
15. *Ozark County News*, June 11, 1891.
16. *Ash Grove Commonwealth*, July 6, 1893.
17. Ibid.
18. *Ash Grove Commonwealth*, July 7, 1892.

19. *Ash Grove Commonwealth,* July 7, 1892.
20. *Stone County Oracle,* June 30, 1904.
21. *Ash Grove Commonwealth,* July 6, 1893.
22. *Douglas County Herald,* July 6, 1903.
23. *Ash Grove Commonwealth,* July 7, 1898.
24. *Cassville Republican,* July 11, 1895.
25. *Douglas County Herald,* July 6, 1905.
26. *Ash Grove Commonwealth,* June 20, 1895.
27. *Houston Herald,* July 11, 1901.
28. *Ash Grove Commonwealth,* July 6, 1893.
29. Ibid.
30. *Marshfield Chronicle,* July 14, 1892.
31. *Ash Grove Commonwealth,* July 6, 1905.
32. *West Plains Journal,* July 16, 1896.
33. *Ash Grove Commonwealth,* July 7, 1892.
34. *Douglas County Herald,* July 7, 1910.
35. *Ash Grove Commonwealth,* July 6, 1893.
36. *Christian County Republican,* June 27, 1901.
37. *Douglas County Herald,* July 7, 1892.
38. *Christian County Republican,* July 10, 1902.
39. *Ash Grove Commonwealth,* July 6, 1893.
40. *Stone County Oracle,* July 18, 1902.
41. *Christian County Republican,* June 27, 1901.
42. George Floy Watter, *History of Webster County, 1855 to 1955,* p. 43.
43. *Ozark County News,* June 2, 1898.
44. *Ash Grove Commonwealth,* May 26, 1892.
45. *Ozark County News,* May 31, 1894.
46. *Ash Grove Commonwealth,* May 26, 1892.
47. *Taney County Republican,* June 1, 1899.
48. Ibid.
49. *Ozark County News,* May 31, 1894.
50. *Ash Grove Commonwealth,* September 8, 1892.
51. *Taney County Republican,* July 23, 1896.
52. *Lawrence Chieftain,* August 25, 1887.
53. *Houston Herald,* August 23, 1906.
54. *Houston Herald,* August 26, 1909.
55. *Lawrence Chieftain,* August 16, 1894.
56. *Lawrence Chieftain,* August 28, 1880.
57. *Stone County Oracle,* August 21, 1902.
58. *Stone County Oracle,* August 14, 1902.
59. *Lawrence Chieftain,* August 26, 1896.
60. *Taney County Republican,* August 18, 1898.
61. *Taney County Republican,* September 17, 1903.

62. *Taney County Republican,* August 27, 1903.
63. *Douglas County Herald,* September 12, 1901.
64. *Cassville Republican,* August 12, 1897.
65. *Cassville Republican,* August 9, 1894.
66. *Ash Grove Commonwealth,* September 8, 1892.
67. *Douglas County Herald,* September 12, 1901.
68. *Douglas County Herald,* September 12, 1901.
69. *West Plains Journal,* August 18, 1898.
70. *West Plains Journal,* August 20, 1896.
71. *West Plains Journal,* August 18, 1898.
72. *Ozark County News,* June 28, 1894.
73. *West Plains Journal,* August 25, 1898.
74. *West Plains Journal,* July 22, 1897.
75. *Stone County Oracle,* July 24, 1907.
76. *Douglas County Herald,* July 15, 1897.
77. *Ash Grove Commonwealth,* July 23, 1903.
78. *Ash Grove Commonwealth,* July 16, 1903.
79. *Lawrence Chieftain,* August 16, 1894.

CHAPTER 8

1. *Cassville Republican,* October 1, 1896.
2. *Ash Grove Commonwealth,* September 20, 1894.
3. *Douglas County Herald,* August 4, 1892.
4. *Douglas County Herald,* September 7, 1888.
5. *Taney County Republican,* October 25, 1899.
6. *Houston Herald,* October 16, 1902.
7. *Ash Grove Commonwealth,* September 29, 1904.
8. *Ash Grove Commonwealth,* October 25, 1894.
9. *Ash Grove Commonwealth,* October 25, 1894.
10. *Cassville Republican,* September 3, 1896.
11. *Cassville Republican,* October 22, 1896.
12. *Taney County Republican,* August 6, 1896.
13. *Cassville Republican,* September 24, 1896.
14. *Douglas County Herald,* November 8, 1888.
15. *Douglas County Herald,* August 25, 1892.
16. *Marshfield Chronicle,* August 4, 1902.
17. Ibid.
18. *Marshfield Chronicle,* June 22, 1897.
19. *Ozark County News,* June 25, 1891.
20. *Douglas County Herald,* July 25, 1901.
21. *Houston Herald,* July 16, 1903.

22. *Douglas County Herald,* July 30, 1903.
23. *Houston Herald,* August 4, 1904.
24. *Stone County Oracle,* September 29, 1904.
25. *Houston Herald,* February 9, 1905.
26. *Ozark County News,* August 29, 1895; reprinted from *Salem* (Ark.) *Banner,* n.d.
27. *Ash Grove Commonwealth,* May 31, 1906.
28. *Ash Grove Commonwealth,* September 26, 1907.
29. *Cassville Republican,* February 18, 1897.
30. *Marshfield Chronicle,* March 24, 1892.
31. *Ozark County News,* March 14, 1895.
32. *West Plains Journal,* August 15, 1901.
33. George Floy Watter, *History of Webster County,* p. 40.
34. May Kennedy McCord, "Our Ozark Christmas Days of Long Ago," *Ozarks Mountaineer,* December, 1953, p. 10.
35. *Christian County Republican,* January, 1903.
36. *Ozark County News,* December 28, 1893.
37. *Lawrence Chieftain,* December 30, 1880.
38. *Ozark County News,* December 28, 1893.
39. *Ash Grove Commonwealth,* December 24, 1908.
40. *West Plains Journal,* January 1, 1903.
41. Ibid.
42. *Cassville Republican,* December 29, 1898.
43. *Christian County Republican,* January 1, 1903.
44. *Cassville Republican,* December 31, 1896.
45. *Ash Grove Commonwealth,* December 2, 1891.
46. *West Plains Journal,* January 1, 1903.
47. *Ash Grove Commonwealth,* December 31, 1891.
48. *Christian County Republican,* January 8, 1903.
49. *West Plains Journal,* January 11, 1900.
50. *Cassville Republican,* December 31, 1896.
51. *Houston Herald,* January 5, 1905.
52. *Ash Grove Commonwealth,* January 2, 1890.
53. *Ozark County News,* December 28, 1893.
54. *Douglas County Herald,* December 31, 1897.
55. *Christian County Republican,* January 1, 1903.
56. *Douglas County Herald,* September 26, 1889.
57. *Ash Grove Commonwealth,* November 5, 1891.
58. *Ozark County News,* February 24, 1898.
59. *Christian County Republican,* April 17, 1902.
60. *Cassville Republican,* May 6, 1897.
61. *Ash Grove Commonwealth,* November 13, 1890.
62. Hollis Lee White, "Occasions for Speech Making in Missouri,

1880–1890" (master's thesis, University of Missouri, 1946), p. 117.

63. *Christian County Republican*, August 16, 1899.

64. W. E. Curry, *A Reminiscent History of Douglas County, Missouri, 1857-1957*, p. 15.

65. *Douglas County Herald*, January 28, 1897; February 4, 1897.

66. *Lawrence Chieftain*, April 21, 1887.

67. Collier, "Ozark and Vicinity in the 19th Century," p. 24.

68. *Houston Herald*, August 23, 1906.

CHAPTER 9

1. *Cassville Republican*, October 15, 1896.

2. *Cassville Republican*, January 11, 1900.

3. *Cassville Republican*, February 25, 1897.

4. *Ash Grove Commonwealth*, September 2, 1886.

5. *Douglas County Herald*, June 16, 1887.

6. *Houston Herald*, December 8, 1898.

7. *Cassville Republican*, February 28, 1895.

8. *Ash Grove Commonwealth*, December 4, 1890.

9. *Houston Herald*, October 9, 1902.

10. *Douglas County Herald*, January 24, 1901.

11. *Ash Grove Commonwealth*, January 23, 1890.

12. *Ash Grove Commonwealth*, December 17, 1908.

13. *Ash Grove Commonwealth*, June 28, 1906.

14. *Taney County Republican*, June 1, 1899.

15. *Douglas County Herald*, December 29, 1892.

16. *Cassville Republican*, October 17, 1898.

17. *Douglas County Herald*, January 27, 1898.

18. Ibid.

19. *Taney County Republican*, March 23, 1899.

20. *Taney County Republican*, February 15, 1900.

21. *Marshfield Chronicle*, January 19, 1903.

22. *Ozark County News*, August 30, 1894.

23. *Cassville Republican*, September 5, 1895.

24. *Ash Grove Commonwealth*, October 6, 1892.

25. *Cassville Republican*, October 27, 1898.

26. *Stone County Oracle*, December 7, 1905.

27. *Cassville Republican*, October 28, 1897.

28. *Houston Herald*, March 10, 1908.

29. *Houston Herald*, June 25, 1908.

30. *Cassville Republican*, December 17, 1896.

31. *Cassville Republican*, December 10, 1896.

32. *Ash Grove Commonwealth*, October 17, 1895.

33. *Ash Grove Commonwealth,* January 12, 1888.
34. *Ozark County News,* September 1, 1898.
35. *Ash Grove Commonwealth,* December 15, 1887.
36. *Ash Grove Commonwealth,* January 31, 1889.
37. *Ash Grove Commonwealth,* October 14, 1886.
38. *Ash Grove Commonwealth,* April 27, 1893.
39. *Ash Grove Commonwealth,* January 31, 1895.
40. *Ash Grove Commonwealth,* October 8, 1891.
41. *Cassville Republican,* January 3, 1895.
42. *Douglas County Herald,* November 10, 1898.
43. *Taney County Republican,* October 19, 1899.
44. *Ozark County News,* August 12, 1897.
45. *Ash Grove Commonwealth,* August 12, 1897.
46. *Ash Grove Commonwealth,* December 13, 1906.
47. *Stone County Oracle,* September 3, 1903; reprinted from the *Aurora Advertiser,* n.d.
48. *Taney County Republican,* October 25, 1899.
49. *Ozark County News,* November 21, 1901.
50. *Ozark County News,* November 21, 1901.
51. *Ash Grove Commonwealth,* February 18, 1892.
52. *Lawrence Chieftain,* September 9, 1886.
53. *Cassville Republican,* May 16, 1895.

Bibliography

BOOKS, ARTICLES, DISSERTATIONS, AND THESES

Ashley, George T. *Reminiscences of a Circuit Rider.* Los Angeles: New Method Printing Co., 1941.

Ayers, Artie. *Traces of Silver.* Reeds Spring, Mo.: Ozark Mountain Country Historical Preservation Society, 1982.

Barker, Mrs. Catherine. *Yesterday Today: Life in the Ozarks.* Caldwell, Idaho: Caxton Printers, 1941.

Bidstrup, Dudley June. "Public Speaking in Missouri, 1840–1860." Master's thesis, University of Missouri, 1940.

Bowen, Elbert R. *Theatrical Entertainments in Rural Missouri Before the Civil War.* Columbia: University of Missouri Press, 1959.

Britton, Wiley. *Pioneer Life in Southwest Missouri.* Kansas City, Mo.: Smith-Grieves Co., 1929.

Broadfoot, Lennie L. *Pioneers of the Ozarks.* Caldwell, Idaho: Caxton Printers, 1944.

Brown, Miriam K. *The Story of Peirce City, Missouri, 1870–1970.* Cassville, Mo.: Litho Printers, 1970.

Christian County: Its First 100 Years. Ozark, Mo.: Christian County Centennial, Inc., 1959.

Clifton, Mary Louise. "Folk Arts of the Ozarks." Master's thesis, University of California at Los Angeles, 1941.

Collier, William Neville. "Ozark and Vicinity in the 19th Century." Privately mimeographed, 1946.

Corchran, Robert, and Michael Luster. *For Love and for Money: The Writings of Vance Randolph, and Annotated Bibliography.* Batesville, Ark.: Arkansas College Folklore Archives Publications, 1979.

Cralle, Walter O. "Social Change and Isolation in the Ozark Region of Missouri." Ph.D. diss., University of Minnesota, 1934.

Curry, J. W. *A Reminiscent History of Douglas County, Missouri, 1857-1957.* Springfield, Mo.: Elkins-Swyers Co., 1957.

Edom, Cliff, and Vi Edom, comp. and eds. *Twice Told Tales and an Ozark Photo Album.* Republic, Mo.: Western Printing Co., 1983.

England, Nerva Brock. *The History of Barry County, Missouri.* Pittsburg, Kan.: Pittcraft, Inc., 1965.

Fairbanks, Jonathan, and Clyde Edwin Tuck. *Past and Present of Greene County, Missouri.* 2 vols. Indianapolis: A. W. Bowen and Co., 1915.

Gerlach, Russel L. *Immigrants in the Ozarks.* Columbia: University of Missouri Press, 1976.

Godsey, Helen, and Townsend Godsey. *These Were the Last.* Branson, Mo.: Ozarks Mountaineer, 1977.

Hall, Leonard. *Stars Upstream.* Columbia: University of Missouri Press, 1958.

History of Laclede, Camden, Dallas, Webster, Wright, Texas, Pulaski, Phelps, and Dent Counties, Missouri. Chicago: Goodspeed Publishing Co., 1889.

History of Lawrence County, Missouri. 3 vols. Springfield, Mo.: Interstate Historical Society, 1917.

History of Newton, Lawrence, Barry, and McDonald Counties, Missouri. Chicago: Goodspeed Publishing Co., 1888.

Hogue, Wayman. *Back Yonder.* New York: Minton, Balch and Co., 1932.

Howard, Guy. *Walkin' Preacher of the Ozarks.* New York: Grosset and Dunlap, 1944.

Ingenthron, Elmo. *Indians of the Ozark Plateau.* Point Lookout, Mo.: School of the Ozarks Press, 1970.

———. *The Land of Taney: A History of an Ozark Commonwealth.* Point Lookout, Mo.: School of the Ozarks Press, 1974.

Johns, Paul W. *Unto These Hills.* Nixa, Mo.: Bilyeu-Johns Enterprises, 1980.

King, Katherine Lillian. "Public Speaking in Missouri in 1880." Master's thesis, University of Missouri, 1943.

Langworthy, Irene. "The Ozarker and His Interpreters." Master's thesis, University of Missouri, 1939.

Lawrence County, Missouri, 1845-1970 — A Brief History. Mount Vernon, Mo.: Lawrence County Historical Society, 1970.

Lawrence County, Missouri, History. Mount Vernon, Mo.: Lawrence County Historical Society, 1974.

Lively, C. E., and C. L. Gregory. *Rural Social Areas in Missouri.* University of Missouri, College of Agriculture, Agriculture Experiment Station, Research Bulletin 305, August, 1939.

McCall, Edith. *English Village in the Ozarks.* Hollister, Mo.: Tri Lakes Press, 1969.

McCord, May Kennedy. "Our Ozark Christmas Days of Long Ago." *Ozarks Mountaineer,* December, 1953, p. 10.

McCurdy, Francis Lee. "Orators of the Pioneer Period of Missouri." Ph.D. diss., University of Missouri, 1957.

McCurdy, Frances Lee. *Stump, Bar, and Pulpit: Speechmaking on the Missouri Frontier.* Columbia: University of Missouri Press, 1969.

Mahnkey, Douglas. *Bright Glowed the Hills.* Point Lookout, Mo.: School of the Ozarks Press, 1968.

―――. *Hill and Holler Stories.* Point Lookout, Mo.: School of the Ozarks Press, 1975.

Marx, Milton, *The Enjoyment of Drama.* New York: Appleton-Century-Crofts, 1940.

Massey, Ellen Gray, ed. *Bittersweet Country.* Garden City, N.Y.: Anchor Press, Doubleday, 1978.

Matthews, Norval M. *The Promise Land: A Story About Ozark Mountains and the Early Settlers of Southwest Missouri.* Point Lookout, Mo.: School of the Ozarks Press, 1974.

Meyer, Duane. *The Heritage of Missouri — A History.* 3d ed. Saint Louis, Mo.: River City Publishers, Ltd., 1982.

Minick, Roger. *Hills of Home: The Rural Ozarks.* San Francisco: Scrimshaw Press, 1975.

Moser, Arthur Paul, comp. "A Directory of Towns, Villages, and Hamlets of Missouri." Vols. 1–9. Typescript, 1981.

255

Nelson, Joseph. *Backwoods Teacher.* New York: J. B. Lippincott Co., 1949.

Official Manual, State of Missouri. Jefferson City: Office of the Secretary of State, 1893/94–1911/12.

Our Heritage in Story and Picture, 1881–1981. Purdy, Mo.: Centennial Circulation Service, Inc., 1981.

The Ozark Region: Its History and Its People. Springfield, Mo.: Interstate Historical Society, 1917.

Packer, Edith. Unpublished diary, 1912–13.

Park, Hugh, ed. *Schoolcraft in the Ozarks.* Reprinted from Henry Schoolcraft. *Journal of a Tour into the Interior of Missouri and Arkansas.* Van Buren, Ark.: Press Argus Printing, 1955.

Pipes, G. H. *Strange Customs of the Ozark Hillbillies.* New York: Hobson Book Press, 1947.

Rafferty, Milton D. *Historical Atlas of Missouri.* Norman: University of Oklahoma Press, 1982.

———. *The Ozarks: Land and Life.* University of Oklahoma Press, 1980.

Randolph, Vance, ed. *An Ozark Anthology.* Caldwell, Idaho: Caxton Printers, 1944.

———. *Ozark Folklore: A Bibliography.* Bloomington: Indiana University Research Center for the Language Sciences, 1972.

———. *Ozark Mountain Folks.* New York: Vanguard Press, 1932.

———. *The Ozarks: An American Survival of Primitive Society.* New York: Vanguard Press, 1931.

———. *Who Blowed Up the Church House?* New York: Columbia University Press, 1952.

———, and George P. Wilson. *Down in the Holler: A Gallery of Ozark Folk Speech.* Norman: University of Oklahoma Press, 1953; reprint, 1979.

Rayburn, Otto Ernest. *Forty Years in the Ozarks: An Autobiography.* Eureka Springs, Ark.: Ozark Guide Press, 1957.

———. *Ozark Country.* New York: Duell, Sloan and Pearce, 1941.

———. "Ozark Folk Encyclopedia." Unpublished work edited by the author. Eureka Springs, Ark., n.d.

A Reminiscent History of the Ozark Region. Chicago: Goodspeed Brothers, 1894. Reprinted, Easley, S.C.: Southern Historical Press, 1978.

Rhoades, Richard, and the editors of Time-Life Books. *The

Ozarks. New York: Time-Life Books, 1974.

Sauer, Carol O. *The Geography of the Ozark Highland of Missouri.* Chicago: University of Chicago Press, 1920.

Schmitt, Alan, and Elizabeth Williams. *City of Seven Hills: A Pictorial History of Ash Grove, Missouri.* Privately printed, 1982.

Sechler, Earl Truman. *Our Religious Heritage: Church History of the Ozarks, 1806-1906.* Springfield, Mo.: Westport Press, 1961.

Shoemaker, Mrs. J. W. *Delsartean Pantomimes.* Philadelphia: Penn Publishing Co., 1891.

Thomas, Jean. *Devil's Ditties.* Chicago: W. Wilbur Hatfield, 1931.

Upton, Lucille Morris. *The Bald Knobbers.* Point Lookout, Mo.: School of the Ozarks Press, 1939.

Watter, George Floy. *History of Webster County, 1855 to 1955.* Springfield, Mo.: Roberts and Sutter, 1955.

Weisberger, Bernard A. *They Gathered at the River.* Boston: Little, Brown and Co., 1958.

West, William Francis, Jr. "The Legitimate Theatre in Rural Missouri from the Beginning of the Civil War Through 1872." Ph.D. diss., University of Missouri, 1964.

White, Hollis Lee. "Occasions for Speech Making in Missouri, 1880–1890." Master's thesis, University of Missouri, 1946.

Willard: From Prairie to Present. Text by Frank Farmer. Willard, Mo.: Willard Bicentennial Heritage Committee, 1976.

Williams, Miller, ed. *Ozark, Ozark: A Hillside Reader.* Columbia: University of Missouri Press, 1981.

Williams, Walter, ed. *The State of Missouri: The Missouri Commission to the Louisiana Purchase Exposition.* Columbia, Mo.: Press of E. W. Stephens, 1904.

Wilson, Charles Morrow. *Backwoods America.* Chapel Hill: University of North Carolina Press, 1935.

———. *The Bodacious Ozarks.* New York: Hastings House, 1959.

Wright, Edward. *A Primer for Playgoers.* Englewood Cliffs, N.J.: Prentice-Hall, 1958.

NEWSPAPERS

Ash Grove Commonwealth, January 2, 1886–December 31, 1910.
Cassville Republican, March 20, 1890–December 31, 1910.

Christian County Republican, November 24, 1898–December 30, 1907; February, 1909–December 31, 1910.

Douglas County Herald, March 10, 1887–December 31, 1910.

Houston Herald, December 1, 1898–December 28, 1899; January 17, 1901–December 31, 1910; June 9, 1949.

Lawrence Chieftain, June 24, 1880–January 20, 1898; January 26, 1899–December 31, 1910.

Mansfield Mail, December 2, 1898–October 28, 1899; January 22, 1904–December 29, 1906.

Marshfield Chronicle, January 2, 1885–October 12, 1893; October 26, 1895–December 31, 1910.

Ozark County News, March 3, 1887–November 24, 1887; January 5, 1889–October 31, 1890; January 3, 1891–December 28, 1899; January 3, 1901–December 29, 1904.

Stone County Oracle, May 3, 1902–December 31, 1910.

Taney County Republican, November 28, 1895–November 19, 1896; November 25, 1897–December 31, 1910.

West Plains Journal, January 2, 1896–December 31, 1910.

PERSONAL INTERVIEWS

Albers, Pansey Powell (Reeds Spring, Stone County).

Allen, Chick (Stone County).

Baker, Walter (Springfield, Greene County).

Bell, Mrs. Queen (Mansfield, Wright County).

Brazeal, John (Springfield, Greene County).

Brown, Mr. and Mrs. E. T. (Gainesville, Ozark County).

Bush, Aunt Man (Reeds Spring, Stone County).

Cranfield, Uncle Joe (Kissee Mills, Taney County).

Davis, Mrs. Docia (Mansfield, Wright County).

Denny, Grover (Cabool, Texas County).

Dunlap, Mrs. Eva (Ash Grove, Greene County).

Fitzner, Mrs. Marjory Powell (Reeds Spring, Stone County).

Hair, Mrs. Earnest (Hurley, Stone County).

Hamilton, Mr. and Mrs. H. A. (Mountain Grove, Wright County).

"Hensley, Squig" (Gainesville, Ozark County).

Hibbard, Claude (Ava, Douglas County).

Hogard, Tom (Gainesville, Ozark County).

Light, the Reverend F. J. (Mountain Grove, Wright County).

Lucas, Mrs. Porter (Crane, Stone County).
McCord, Mrs. May Kennedy (Springfield, Greene County).
McKinley, Joe (Springfield, Greene County).
McPherson, Mr. and Mrs. Frank (Ash Grove, Greene County).
May, Fred (Galena, Stone County).
Miller, Sam (Fordland, Webster County).
Morrill, Oscar (Notch Post Office, Stone County).
Pipes, Gerald (Reeds Spring, Stone County).
Powell, Miss Hazel (Reeds Spring, Stone County).
Powell, Walker (Reeds Spring, Stone County).
Renfro, Ambrose (West Plains, Howell County).
Rideout, Tom (West Plains, Howell County).
Roy, Ezra (Mansfield, Wright County).
Roy, Jason (Mansfield, Wright County).
Scott, Rufe (Galena, Stone County).
Steele, Fred (Hurley, Stone County).
Vandiver, E. H. (West Plains, Howell County).
Wallis, L. O. (Ebenezer, Greene County).
Weaver, Charlie (Springfield, Greene County).
Yoeman, Emmett (Ava, Douglas County).

Index